当代世界建筑经典精选（7）

SOM 建筑设计事务所

SKIDMORE, OWINGS & MERRILL

Selected and Current Works

世界图书出版公司

北京·广州·上海·西安

当代世界建筑经典精选(7)

SOM 建筑设计事务所

SKIDMORE, OWINGS & MERRILL

Selected and Current Works

世界图书出版公司

北京·广州·上海·西安

First published in Australia in 1995
Second Edition 1997
The Images Publishing Group Pty Ltd
ACN 059 734 431
6 Bastow Place, Mulgrave, Victoria, 3170
Telephone (61 3) 9561 5544 Facsimile (61 3) 9561 4860

National Library of Australia Cataloguing-in-Publication Data

Skidmore, Owings & Merrill LLP.
SOM: selected and current works.

2nd ed.
Bibliography.
Includes index.
ISBN 1 875498 66 4.
Master Architect Series ISSN 1320 7253

1. Skidmore, Owings & Merrill LLP 2. Architecture, American.
3. Architecture, Modern—20th century—United States.
4. Architects—United States. I. Title. II. Title: Skidmore, Owings & Merrill LLP. (Series: Master architect series).

720.92

Edited by Stephen Dobney
Designed by The Graphic Image Studio Pty Ltd,
Mulgrave, Australia

Printed by Everbest Printing in Nansha, Panyu, China

Contents

Selected and Current Works

Design for Commercial Use

Contents Continued

Introduction

SOM 1984–1994

By Joan Ockman

As one of the largest and most prestigious architectural firms in the world, Skidmore, Owings & Merrill (SOM) has for half a century set the standard of American corporate design practice. In the decades after World War II, SOM's name was synonymous with some of the most illustrious examples of International Style architecture. In more recent years, as modernist orthodoxy has been revised, the firm has continued to occupy the forefront of a field that has become both more aesthetically diverse and more geographically fluid, presiding over a transition from International Style to global practice.

A relative index of this change is provided by the series of monographs SOM has published over the last four decades. Of the projects included in the volumes for the 1950s and 1960s, those outside the United States accounted for barely 10 percent of the total. In the volume for the 1970s, when major commissions came from the Middle East, this number doubled. In the current volume, almost half the work is in foreign countries.

Following the new inroads of capital accumulation, SOM's projects today range the globe from Manhattan to Moscow, Chicago to Shanghai, London and Berlin to Kuala Lumpur, Jakarta, and Ho Chi Minh City. It is a historical irony that the modernist aspiration to a world architecture is being realized within a "postmodernist" cultural climate. It is also remarkable that precisely the areas that have witnessed America's greatest international conflicts—Germany, Japan, Russia, China, Korea, Vietnam—are now offering SOM some of its major opportunities.

SOM's experiences exemplify the dramatic changes that have affected the profession in the last decade. While the Cold War's end and new computer and telecommunications technologies promised to transform practice in far-reaching ways, a recession in 1987 created more immediate uncertainties for architects, halting an ambitious array of commercial and urban development schemes launched earlier in the decade. In response to these changing conditions, SOM, like other large practices at home and abroad, scaled down and shifted orientation. Long known for its multi-disciplinary in-house structure, which enabled it to offer comprehensive architecture, planning, interior design, and engineering services, the firm now emphasizes flexibility and efficiency. Certain specializations have been eliminated in some of its offices and the use of inter-office teams and outside consultants has increased. This restructuring has been greatly facilitated by the new technologies.

Beyond such organizational changes, the impact of globalization has manifested itself in what is literally a new "world view." In this regard, a project now being designed by SOM for a location in Hawaii but planned as a prototype

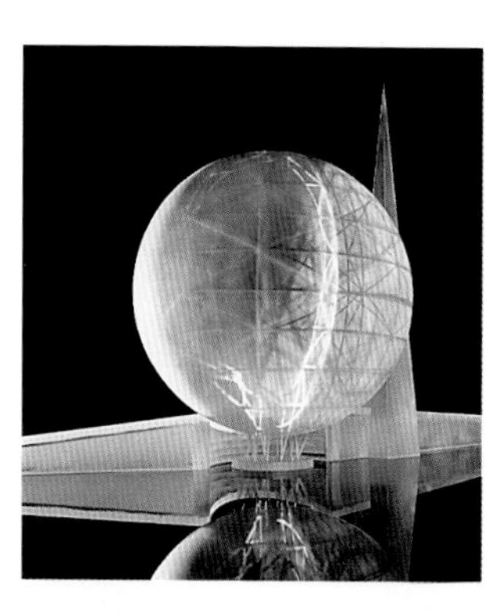

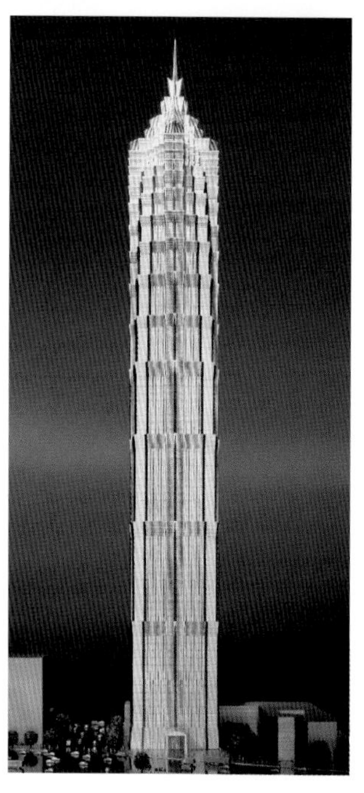

could hardly be more emblematic. An immense spherical entertainment and commercial center, SOM's World Trade Center Prototype centers on a multi-use stadium at the base of a vast atrium and is linked by a bridge to a pyramidal elevator tower. The curved surfaces of its inner and outer shells double as giant telescreens onto which a computer maps world events.

The architectural dream of internationalism—initially embodied in the world's fairs of the second half of the 19th century and permeated with utopian idealism in modernist projects like Le Corbusier's Mundaneum—is here animated less by the spirit of progress and production than by that of profit and play. Yet it appears no less optimistic in its wedding of monumental form to high technology. Between the virtuosity of SOM's global theme park and such formal predecessors as Boullée's Cenotaph for Newton with its awesome metaphysical interiority or Leonidov's lightly tethered Lenin Institute aspiring toward the perfect socialist society, a long distance has been traveled. Closer in its vision of the future to the centerpiece Trylon and Perisphere at the 1939 New York World's Fair (a fair which gave SOM early work and helped to boost its career), the sphere for Hawaii combines popular culture, current aesthetic trends, and advanced engineering into an audacious icon of late 20th century capitalism.

No less symbolic of the ambitions associated with an epoch of multi-nationalism and global communications are the tall buildings designed by SOM during the past decade. These high profiles, including a handful of "supertowers" exceeding Sears in Chicago (at this point still on paper), convey the power and affluence of clients that can command so formidable an amount of ground and air space. Apart from the problems of engineering them—above 90 stories they exact increasingly rigorous structural demands—the continuing competition to dominate the skyline raises basic questions of economic feasibility and environmental impact.

Contemporary clients, however, often tend to be more concerned with matters of symbolic representation. Should a skyscraper in Shanghai look like buildings in China or Chicago? The answer depends on the specific conjunction of economics, politics, and culture. A supertower in a developing area of the world is above all a lightning rod for investment, a hypodermic to inject capital into the economy. The subject of regionalism has been rather sentimentally broached by architects in the last decade as a matter of preserving cultural differences. But the transition from donkey cart to supersonic jet is not necessarily palliated by the recall of familiar forms. From a more realistic (and less patronizing) standpoint, Western architects are called to build in developing countries today for two principal

reasons: to provide expertise in handling complex design problems, and to give currency and cultural value to a focal building intended to function as a monumental sign. SOM's buildings tend to fulfill these requirements with great skill.

Ideally the global skyscraper also goes further, offering an eloquent and unsentimentalized response to its context. An admirable example is SOM's National Commercial Bank in Jeddah, Saudi Arabia. When, in the 1930s, the brise-soleil wall was first invented by modern architects, it represented an original solution to building in tropical climates. By the 1950s and 1960s, however, it had largely become a cliché applied reflexively to the universal glass box. In SOM's bank for this desert city, the typology of the high-rise in a harsh environment is entirely rethought. The window wall is abandoned for three stark facades uninterrupted except by multi-story square apertures that filter sunlight and views into elevated courtyards overlooked by glazed interior office walls. The building's pure triangular volume, a minimalist sculpture rising from the flat landscape, comments on the transparent modernist prism, but now translated into the taut planarity of travertine. At the same time, the triangle's reiteration at the scale of the paving patterns and the opulent materiality recall motifs of Arabic culture in a nonliteral way, creating a powerful monument for a sophisticated commercial institution in a traditional national setting.

Similarly but at a smaller scale, another bank building in the Middle East, United Gulf in Manama, Bahrain, uses an elegantly articulated curved plane of sunscreening, aerial cross-walks, and jewel-like detailing to make high technology and monumental form sympathetic to local climate and cultural tradition.

In both instances, SOM was able to carry its design through most stages of development. Such opportunities, however, are becoming rarer as the "architecture" component in commercial buildings is increasingly relegated to the skin, crown, and lobby. This phenomenon is a function of two factors. First, advances in technology have enabled the high-rise's engineering design to be dissociated from its surface, making the tower a structure to be sheathed, a variation on a theme. The computer's integration in the design process fosters this game, multiplying visual possibilities by enabling alternatives to be studied readily. Second, while commercial clients today seek a distinctive image for the most public parts of their buildings, at ground level and on the skyscape, they require maximum flexibility and interchangeability for the rentable space within.

This too represents a significant change for SOM. While the major portion of its work has always been commercial, in the past its primary clients tended to be owner-occupants, hiring SOM to take

the design "from master plan to ashtrays." Today, however, when fewer owners can afford to be burdened by real estate contingencies, clients tend to be speculators seeking a marketable image for generic office space. Often composite and unlocalized entities, they prefer to control costs and risk by decentralizing responsibility, looking to architects especially to confer prestige.

The reputation of SOM's architecture, to its great credit, has always been predicated on the overall building rather than on a superficial "signature." Yet the disjunctiveness of the contemporary design process poses new challenges to an architecture that prides itself on more than facadism. Perhaps the hybrid play of grids, layers, and elements in its recent scheme for the edge of Alameda Park in Mexico City is already a commentary on this condition. This project occurred in context of a collaboration with Frank Gehry and Ricardo Legorreta, in which the three firms designed adjacent towers in a spirited "montage of attractions."

Another unconventional association produced a winning competition entry for Shenzhen International Economic Trade Center. For this 88-story tower and mixed-use development in China, a SOM team traveled to the site to join architects from the state-chartered design institute of the local university; should the project go forward, SOM will continue to work with the same architects. Especially in countries where regulations or politics ensure that a design's execution is entrusted to local architects or consultants, such collaborative arrangements are increasingly routine. Offering unprecedented opportunities for creative dialogue and engagement, the new realities of globalism demand innovative methods and strategies for controlling the design's quality and outcome.

Contradicting the implications of such radical changes—perhaps unconsciously to ward them off—architecture during the last decade exhibited a taste for extremely rich materiality and detailing. SOM has always been known for its superb standards of construction and fabrication, as also demonstrated over the years by the caliber of its interior design, and its recent work remains within this tradition. But in the days when the deity in the details was Miesian, the meaning of refinement seemed to be linked to the industrial precision of metal and glass. Is it an anachronism that the care once taken with expressing steel mullions and spandrels is now also being lavished on cut stone, polished wood, and custom fixtures by artisans working from computer drawings on fast-track contracts?

The revival of craft technique in an age of machine reproduction and urban flux is purely a perquisite of wealth, and yet the rarefied International Style details of Chase Manhattan Bank or Union Carbide, for all their restraint, did not come off the

shelf. If the posh lobbies and extravagant materials typical of 1980s taste conspicuously display their owners' privilege, perhaps it is less a matter of "cultural degeneracy," as Adolf Loos would have had it, than of society's need for greater sensual delight. The Sheraton Palace in San Francisco, an expert restoration of a historic grand hotel, gives evidence of this, as does the several million square feet of interiors designed for Merrill Lynch's corporate headquarters in the World Financial Center in New York City. The argument is all the more persuasive when such pleasures are shared generously with the public.

Indeed, the recent revision of modernist orthodoxy has resulted in a welcome emphasis on highly amenable and representative public space, especially at the scale of the city. Whether this is to ignore or to compensate for the universalizing effects of globalization, it is here that the ambivalence between past and future reveals itself most dramatically within late 20th century architectural practice. The theme is amply illustrated in several ambitious urban redevelopment schemes undertaken by SOM in the last decade, where architectural primacy is subordinated to the enhancement of collective use.

At Canary Wharf in London, a 19th century dock area extensively damaged by World War II bombs, SOM was responsible for the master planning, design guidelines, and four office buildings on the site. While the classicism of the plan and the evocations of Edwardian taste reflect the past two decades' preoccupation with historical forms, SOM has also provided public space that is well designed, beautifully crafted, and up to date (in consultation with a talented landscape firm, Hanna/Olin, Ltd.). The same is true at Rowes Wharf in Boston, where architecture and urban design merge to affirm a past when architectonic monumentality was a civic virtue.

It is no accident that obsolescent 19th century ports and railyards have become a prime focus for such urban regenerations. Their often spectacular sites and the nostalgia that a post-industrial society attaches to industrial ruins make these large-scale former places of production ideally suited for conversion to zones where public and commercial space overlap. SOM's master plans for Riverside South in Manhattan, Mission Bay in San Francisco, the Hanseatic Trade Center in Hamburg, and Osaka Sakai Seaport further illustrate the trend. Riverside South dramatizes the stamina required of architects given the political and economic stakes associated with such sites; the master plan there resulted from over 700 meetings with the client, community boards, and other local interests (actually not so surprising in as contestatory a city as New York).

Probably the outstanding example of public urban design and high-quality architecture built by SOM in the last decade is to be seen at Broadgate, another major redevelopment site in London which again involved reusing a 19th century infrastructure. Here the site's special nature—air rights over Liverpool Street Station rail lines—produced a contextual dialogue of an entirely different sort. The fluid central plaza, also designed with Hanna/Olin, Ltd., effectively mediates between the great arching ironwork shed of the train station and the dark green cage of SOM's Exchange House facing it. The building's extruded steel framework straddles the tracks, incorporating a series of four giant parabolic arches that alliterate the station's roof. It is at once a hybrid structure and a very pure synthesis of the site's levitational energies and flows. The latter arrive at an exquisite threshold in the building's suspended open lobby.

Besides adapting former industrial infrastructures to new usages, SOM's planners and designers have also turned their energies to revitalizing older transportation facilities and designing 21st century systems. The firm's master plan and renovation of historic stations along the Northeast Corridor for the Federal Railroad Administration and Amtrak celebrate train travel in America while also promoting its viability. At the same time, its work at air terminals in major American cities and abroad exploits cutting-edge technology. In Seoul a hangar bracketed on three sides by an eight-story office annex, currently under design for Korean Airlines, displays an innovative use of building systems. Fabricated from standard components, its roof structure consists of convexly and concavely arched steel trusses arranged in a V-formation and supported on three box columns. The huge uninterrupted span poetically alludes to the winged aircraft housed beneath.

From high-tech structures to finely detailed interiors, from superscale office towers to urban place-making, SOM's work traverses an extremely wide range of architectural practice, demonstrating the firm's commitment to the broadest possible provision of design services. While those aspects of its work that most reflect the new impact of globalization and post-industrialism have been emphasized here, the second section of this catalogue illustrates SOM's distinguished handling of museums, religious edifices, courthouses, transitional housing, and other institutional building types. These further reveal its success in providing exceptional levels of public amenity. Also notable within the category of design "for the public" are its carefully executed health, leisure, and service facilities. These sometimes "invisible" environmental interventions help to extend the purview of contemporary design practice.

The programmatic, geographic, and aesthetic diversity inherent in this extensive body of work makes SOM's achievement unique in the field and resistant to generalization. While the firm's collective signature has always stood for the different personalities of its partners and design teams, its architecture in the past could still be placed within the developmental framework of canonical buildings like Lever House and Inland Steel. This is no longer the case. Is it, then, still possible to recognize a SOM building as such? Buildings like Southeast Financial Center in Miami or Citicorp in Long Island City, which elegantly extend the firm's rationalist tradition; the wharf projects for London and Boston with their historicist imagery; or, say, the winning competition project for a mixed-use development at Checkpoint Charlie in Berlin, animated by constructivist kinetics, would seem to represent disparate points on an ideologically embattled aesthetic spectrum. SOM has remained distanced from the vanguard debates that preoccupy journals and architecture schools, yet their traces are reflected in its work, attesting not only to the osmotic relationship of theory to practice in architecture, but also to the firm's bid for currency in a competitive field where taste is strongly shaped by the media and "pluralism" is a current watchword.

At the same time, a building by SOM today is above all a product of a precise "fit" between client, architect, and context. The decision as to whether it is made of granite and marble or glass and steel is an outcome of this relationship, not a matter of ideology. Diversity derives from conscious empiricism rather than willful eclecticism. In this sense, while SOM's variegated recent work departs from post–World War II orthodoxy, it redefines modern architecture's tradition of "problem solving." Architects disdainful of the notion of design as a service profession or made cynical by the complex realities of normative practice may naively imagine their art can be principally a matter of aesthetic theory or bravura form-making. Such attitudes, however intellectually and visually provocative at times, contribute to the profession's lamented marginalization. Ultimately having a far more powerful impact on the future of both the architectural profession and the built environment are the radically new economic and sociocultural forces that are currently reshaping the world.
In this context of change, what continues to distinguish a SOM building is its masterly ability to translate contemporary conditions of practice into an exacting and sophisticated art of building.

Joan Ockman

Director of the Temple Hoyne Buell Center for the Study of American Architecture at Columbia University Graduate School of Architecture, Planning and Preservation.

Selected and Current Works

Design for Commercial Use

Broadgate Development

Design/Completion 1987/1990
London, England
Rosehaugh Stanhope Developments PLC
4,000,000 square feet
Exposed steel frame
Painted metal and granite

The master plan, and architectural and structural engineering design of ten buildings in the Broadgate complex are part of the largest single development in the City of London. Most of the development is built over the platform and railway tracks of Liverpool Street Station. The multi-use complex expands the financial district of London by providing new office space and trading floors, and enhances the surrounding urban district with retail and leisure facilities. Three public squares and the terraces and landscaping of Exchange Square provide a focal point for the complex, with spaces designed for performance and recreation.

The ten buildings are designed in a variety of styles to relate to the City context. Exchange House is a ten-story office block supported on an expressed structural frame that spans the Liverpool Street Station tracks. The buildings facing Bishopsgate present more traditional facades of carved stone and polished granite.

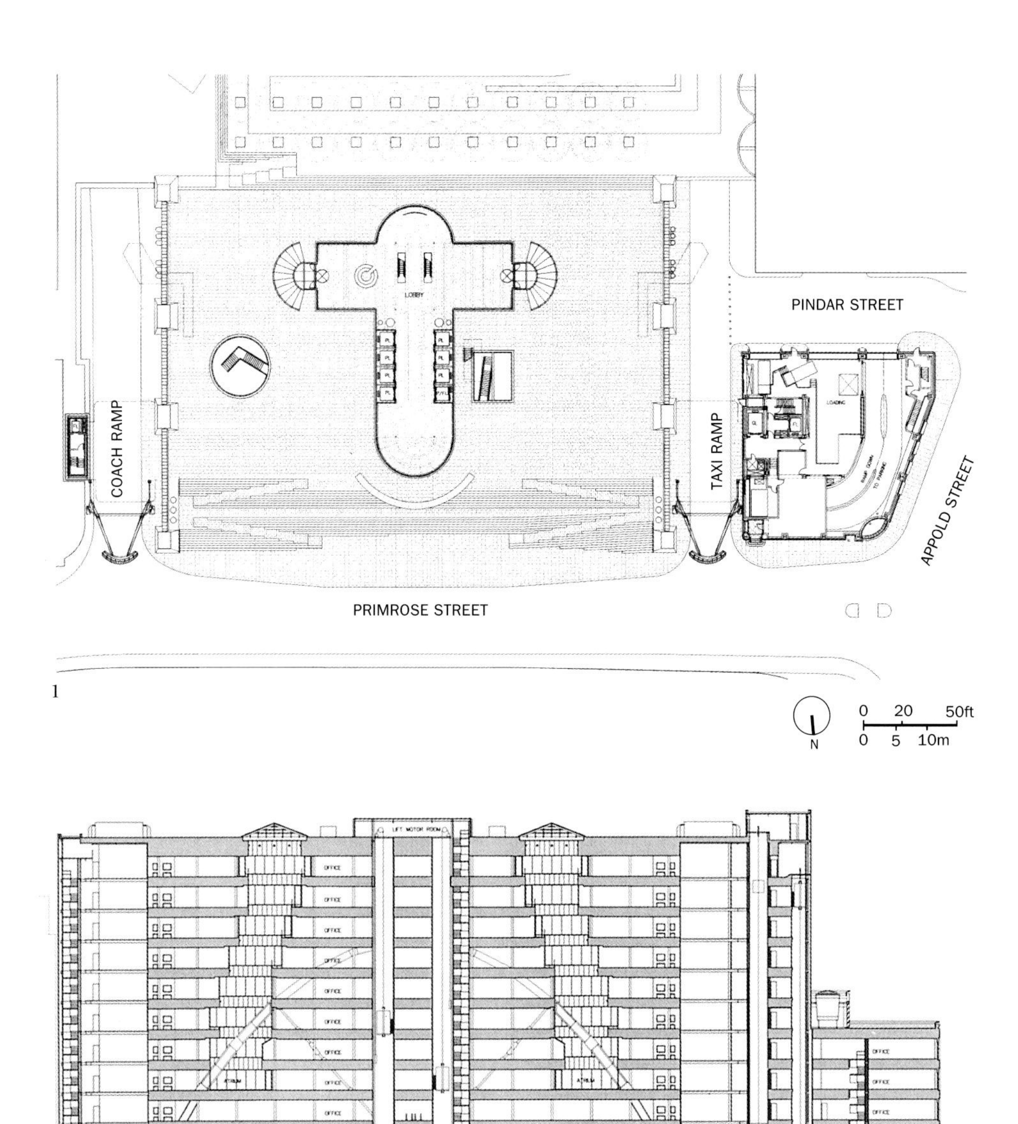

1 Exchange House, plaza-plan level
2 Exchange House, east–west section
3 Aerial view of model

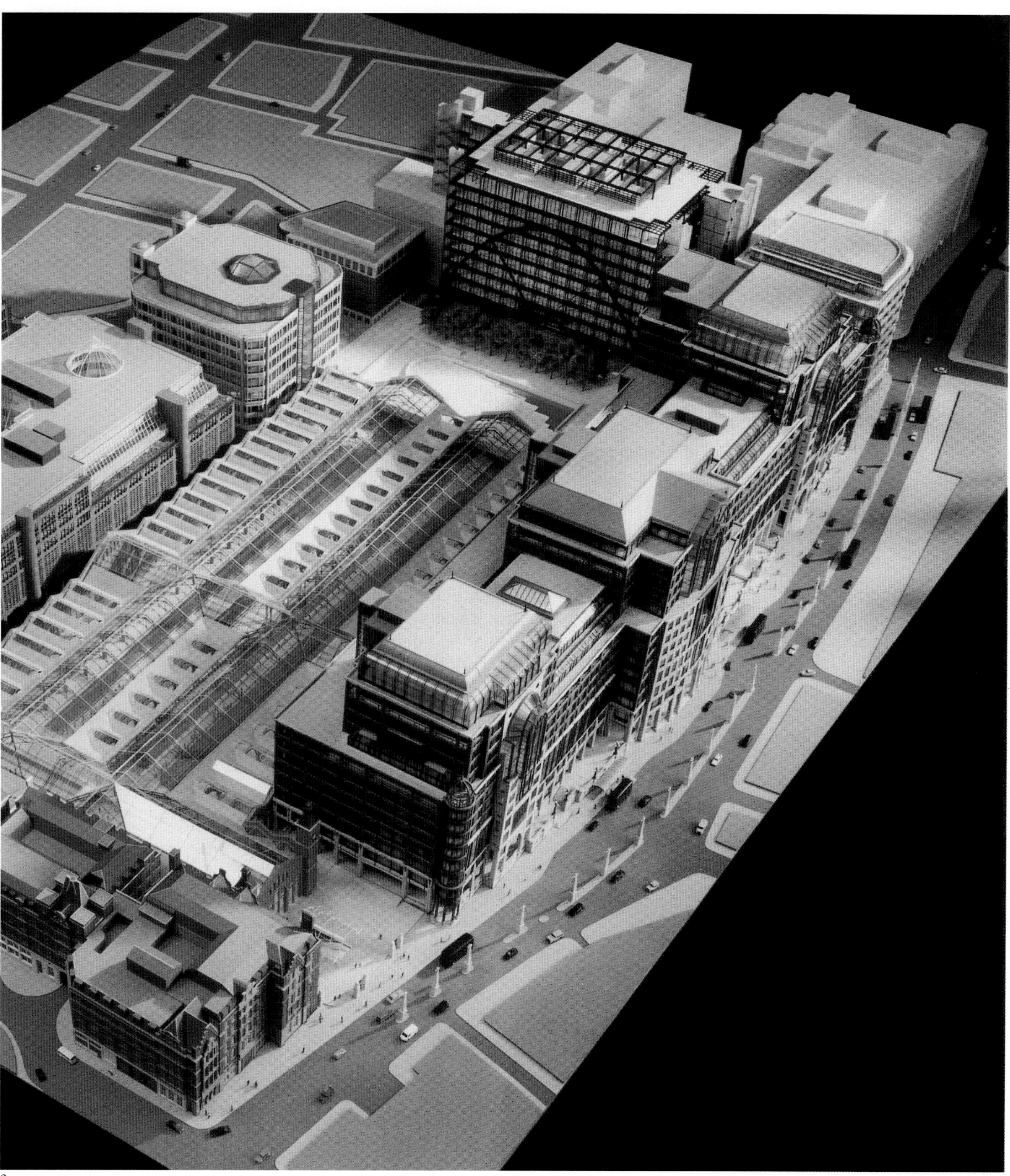

3

4 View of Exchange Square
5 South elevation
6 View of Exchange House

4

5

6

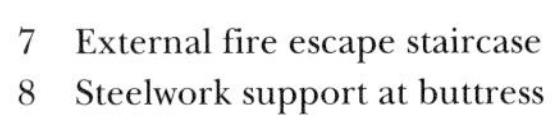
7 External fire escape staircase
8 Steelwork support at buttress

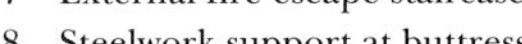

7

8

Ludgate Development

Design/Completion 1990/1992
London, England
Ludgate Properties PLC
390,200 square feet
Steel frame
Limestone, granite, painted metal, stainless steel

A development consisting of three buildings on a very constrained site, the project is situated over the Thameslink railway in a densely populated neighborhood. The height of the development is restricted by the St Paul's Height View Corridor.

A complex civil engineering design realigned railway lines and removed a viaduct and railway bridge, creating a long, narrow site. The buildings are grouped around a new city square and narrow pedestrian walkways characteristic of the district.

The exposed structural steelwork and vertical metallic blue fins of 1 Fleet Place add depth and texture to the elevations. At 10 Fleet Place, the dark granite and steel facade is characterized by the strong vertical elements contrasting with horizontal metal spandrel panels and stainless steel bands.

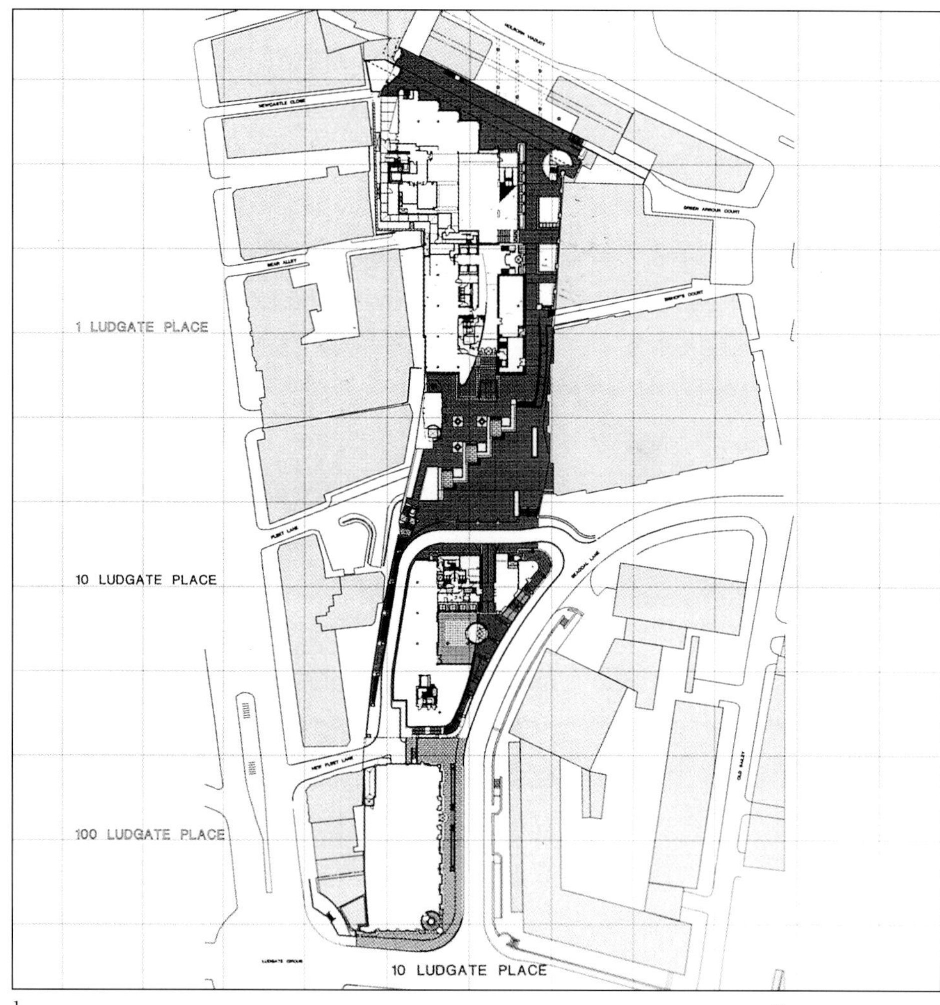

1

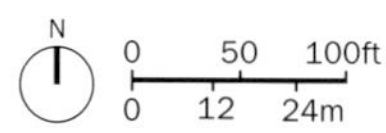

2

1 Site plan
2 10 Fleet Place lobby
3 10 Fleet Place, north-east corner

3

4–8 Exterior wall detail, 10 Fleet Place

4

5

6

7

8

Rowes Wharf

Design/Completion 1982/1987
Boston, Massachusetts, USA
The Beacon Companies
665,000 square feet
Steel frame
Brick and precast concrete cladding

Rowes Wharf accommodates a complex program of offices, condominium residences, restaurants, underground parking, shops, a hotel and health club, and facilities for private and commuter boats in an assemblage of buildings. The development is on a 5.4-acre land and water site, two-thirds of which is devoted to open space. By continuing the urban facade, the building is decisively reconnected with the urban fabric; by following the curve of Atlantic Avenue, the design reinforces the street edge and defines the street as the primary element of the public domain.

Like the existing wharf buildings that characterize Boston Harbor, the new buildings are straightforward, almost industrial in their appearance. The complex is open to those approaching the city by boat, with a central arch that functions as a gateway into the city.

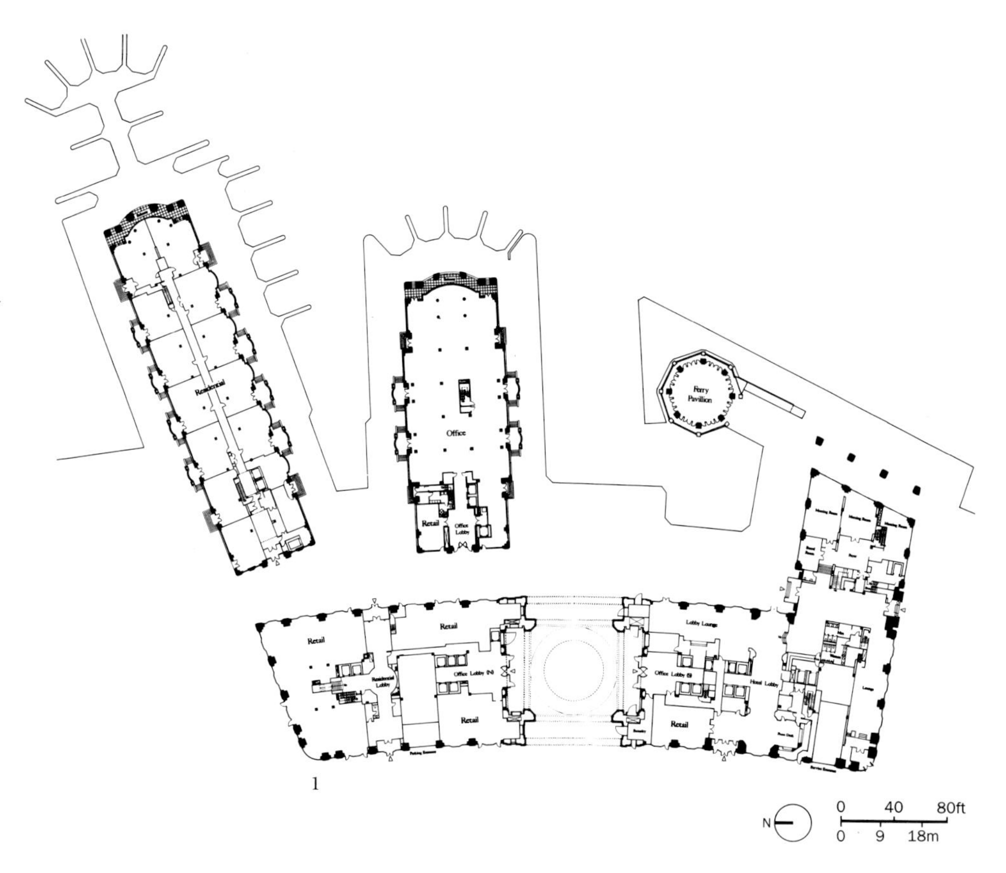

1

2

1 Ground-floor plan
2 View of ferry terminal
3 View from harbor

3

4 View of arch
5 Office entry

4

5

Reconstruccion Urbana Alameda

Design/Completion 1993/1995
Mexico City, Mexico
Reichmann International Mexico
5.2 million square feet (site area: 450,000 square feet)
Steel and concrete
Curtain wall and concrete

Fronting on the Alameda, the park that is the cultural and historical heart of Mexico City, the Urbana Alameda district links the historic center with the Avenida Paseo de la Reforma, the main commercial avenue. The master plan and concept design for the mixed-use development are the first phase of the massive rejuvenation of the city's principal business, hotel and entertainment district that was destroyed by the 1985 earthquake.

Principal elements of the project include a large-scale business complex, comprising five towers and three mid-rise buildings, immediately accessible to the established financial precinct nearby; a luxury hotel facing the Alameda, possibly including serviced apartments; and a major new complex of shops, restaurants, and public spaces for area residents, office workers, and tourists. The project was designed in collaboration with Frank Gehry and Ricardo Legorreta. Its eclectic architectural vocabulary responds to the existing downtown fabric while reinterpreting materials and colors typically used in Mexican building.

1

1 View from park looking east
2 View from park looking west

2

Columbus Center

Design/Completion 1987/
New York, New York, USA
Boston Properties
2,000,000 square feet
Steel and concrete
Brick and granite veneer

Columbus Center, a multi-use complex containing offices, retail, residences, and parking, is to be built at 59th Street and Columbus Circle at the south-west corner of Central Park. This is one of the few remaining sites with an axial relationship to the Manhattan grid. A great sweeping base of retail and offices curves along Columbus Circle supporting three towers designed in the architectural vocabulary of the nearby twin-tower apartment buildings along Central Park West.

The granite base houses a four-story retail element that continues the pedestrian scale of Broadway, graduating to a 21-story office component. Three paired residential towers set back from the office and retail base rise to varying heights, culminating 62 floors above the Circle. A "window" between the towers extends the view corridor west from 59th Street. A grand rotunda is the major public entrance and organizing element for the complex.

1

1 Computer rendering of central towers
2 Perspective view of Columbus Circle

2

CSC Bangkok

Design/Completion 1991/
Bangkok, Thailand
City Realty Company, Ltd
3,280,000 square feet
Cast-in-place concrete
Ceramic tile

Located on the Chao Praya River within the city limits, the mixed-use project comprises three 45-story luxury condominium towers, a 40-story office tower, a 300-room hotel, and community amenities that include retail, restaurants, and health clubs. A clearly defined residential precinct along the waterfront and a commercial/office precinct on the street side characterize the plan for the complex—a largely pedestrian environment with residential components grouped around a large landscaped area and man-made lagoon.

The hotel element links the two precincts via an eight-story portal cut through the building mass. The condominium tower plans accommodate a maximum of three units per floor, with sizes ranging from 3,000 to 4,000 square feet per unit. Facades incorporate brises-soleil and generous exterior terraces. Penthouse units feature outdoor gardens and terraces protected by high roofs for use during the monsoon season.

1

1 Site plan
2 Elevation and floor plan
3 Axonometric drawing of site

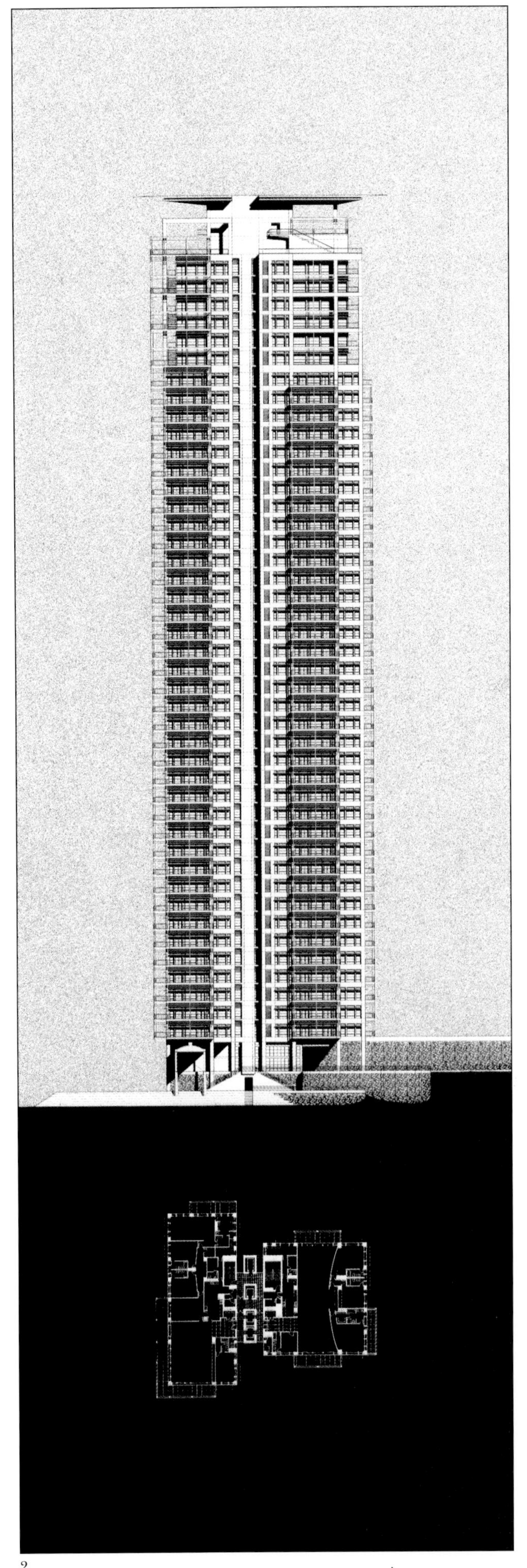

2

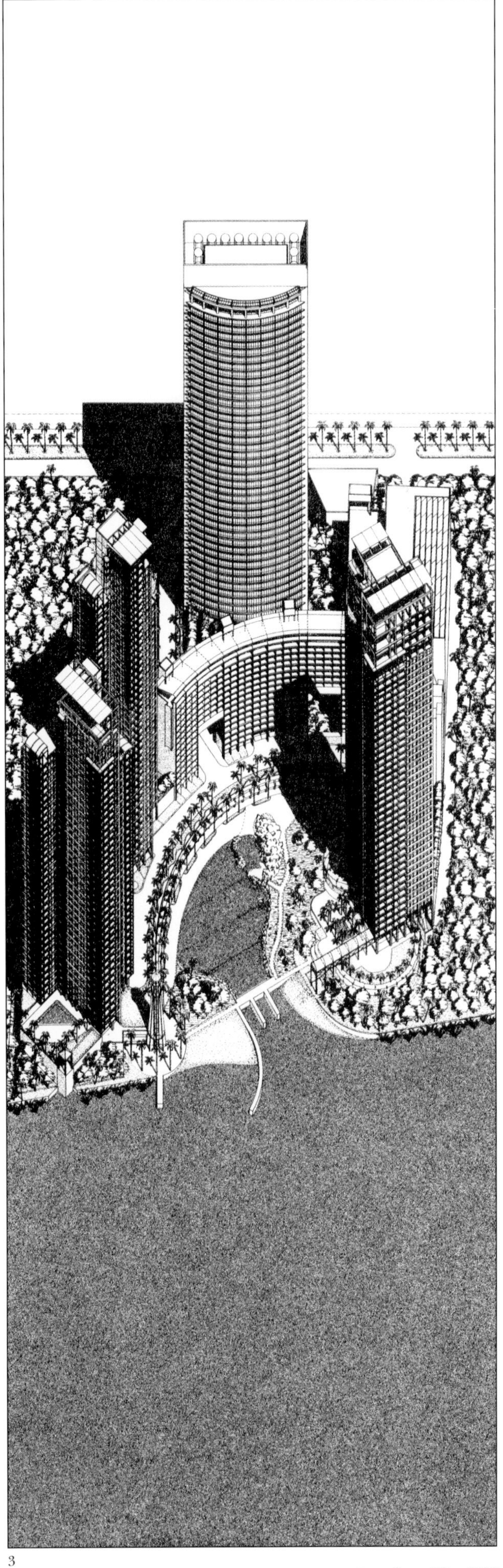

3

Kuningan Persada Master Plan

Design/Completion 1993/1995
Jakarta, Indonesia
P.T. Kuningan Persada
4,306,000 square feet
Steel frame (tower); concrete frame
Glass and metal curtain wall

The rational assignment and integrated support of mixed land uses are at the heart of the Kuningan Persada superblock development master plan. The 70-story (1,480 foot high) office tower, crowned by Jakarta's telecommunications center, is a landmark for the surrounding area. The tallest structure in Jakarta, it incorporates observation decks and restaurants, as well as office space for international and Indonesian businesses seeking a world-class building and the best international communication linkages in the region.

Enhancing public perception of the development as an international business center is the five-star, 32-story hotel that is the gateway to the complex. A 753,500 square foot retail mall completes the land-use program; adjacent to the hotel and linked to the tower, it provides a luxurious shopping environment for residents and tourists alike.

1

1 Aerial view of Kuningan Persada superblock development
2 View of office tower

2

Sentul Raya Square

Design/Completion 1993/1994
Kuala Lumpur, Malaysia
Mayang Sari Development
5,531,000 square feet: 440,000 square feet of sports arena; 200,000 square feet of exhibition space; 425,000 square feet of entertainment space; 40,000 square feet of museum; 111,000 square feet of festival retail space; 110,000 square feet of neighborhood retail; 1,000,000 square feet of hotels (three hotels); 1,000,000 square feet of office space; parking for 6,300 cars

This project is a master plan for a mixed-use development conceived as a major cultural, entertainment and commercial destination, located near the heart of Kuala Lumpur on the site of the historic KTM Railroads. The architectural design approach incorporates several historic buildings and relies on the use of new building materials that are historically accurate.

The entertainment tower marks the location of the square, and is prominent on the skyline. Other components include the entertainment center, a five-story rotunda building for retail and entertainment; the Cultural and Railroad Museum; the retail mall tower, with more retail, a food court, and transportation connections; the arena, a 15,000-seat venue for sports, concerts, and exhibitions, flanked by two exhibition halls; a shopping arcade; and a five-star hotel linked directly to the arena by a second-story bridge.

1

2

1 Arena interior showing boat show
2 Railway museum
3 View of south entry drive
4 Model view

3

4

Shenzhen International Economic Trade Center

Competition/Completion 1992/1997
Shenzhen Special Economic Zone, Guangdong Province, China
Sinotrans Shenzhen Development Co.
3,500,000 square feet
Structural steel frame
Metal, stone, and glass cladding

The Shenzhen International Economic Trade Center is a trade exposition complex organized along a riverfront site, containing large- and small-scale exhibition halls; a hotel, office, and residential tower; two mid-rise residential towers; and a large retail center.

The project is the first large commercial development outside of the central district. As the gateway to a new area, the complex is a landmark and is completely self-sufficient as a destination point. While the 88-story tower dominates the skyline, the street-level retail center is immediately visible from the arrival road. Amenities such as a public garden and "floating" riverfront restaurant make the development a destination for the region.

SOM associated with Shenzhen University Institute of Architectural Design for the design competition, which it won unanimously.

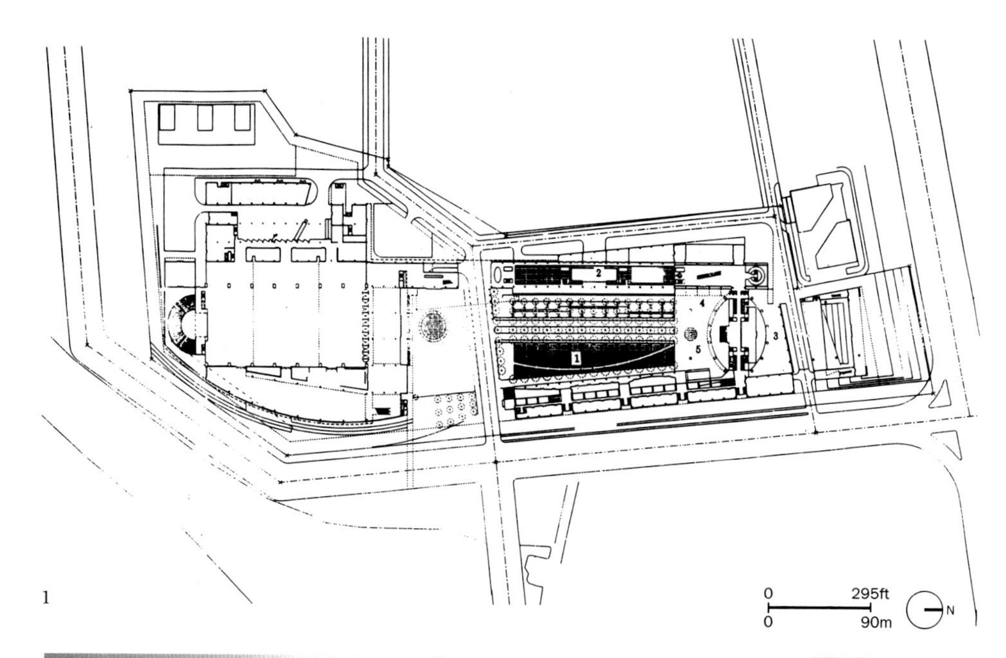

1

2

1 Ground level plan
2 View of exhibition hall
3 View of office/hotel tower

3

Urban Redevelopment Authority, Parcel A Hotel

Competition 1989
Singapore
Wing Tai Holdings Ltd/Straits Properties Ltd
438,800 square feet
Concrete frame structure
Metal, stone, and glass cladding

The project combines a 350-room hotel, a large retail pavilion, and four levels of below-grade parking on a compact and partially developed site on Orchard Road, an area of Singapore well-known for its prestigious hotels and shopping centers. The competition was sponsored by the city's Urban Redevelopment Authority.

The hotel design organizes the guest rooms and circulation around a cylindrical atrium. An innovative radial frame structure supports the tower—a staggered shear/bearing wall system that can be built quickly and cost-effectively with a limited number of components. While the tower gives the hotel a distinctive identity, its low-rise elements, including the retail pavilion, anchor it to the site and link it to the adjoining office building.

1

2

1 View from Orchard Road
2 Interior hotel courtyard
3 View of hotel tower and support facilities

3

Hanseatic Trade Center

Competition 1990
Hamburg, Germany
1,410,100 square feet (site area: 447,750 square feet)

The program for the competition site on the Elbe River includes offices, a 250-room hotel, and retail space, as well as the development of waterfront parks. In organization, massing, and use of materials, the design draws both from the historical architectural context of the waterfront and from contemporary vocabularies, anticipating the trade center's position as a focus for newly established East–West trade relations.

The buildings proposed for the site are of three types: linear courtyard buildings that continue the rhythm and massing of adjacent warehouses; buildings conceived as a series of pavilions; and object buildings that are the culmination and termination of the linear buildings. Central to the complex, the glass and metal-clad hotel echoes neighboring tower forms. Several bridges link the two "arms" of the site, as well as connecting the site to riverbanks on either side.

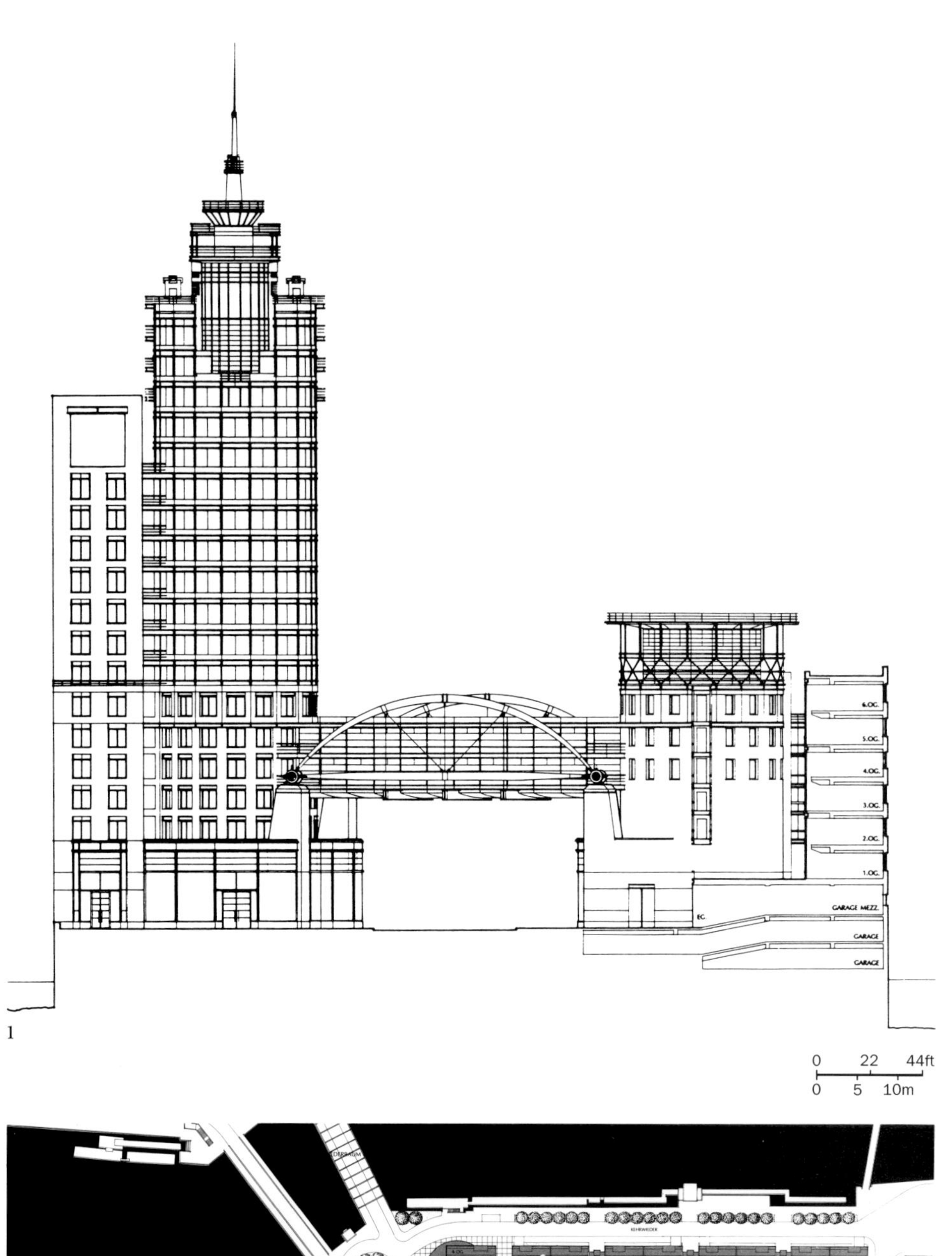

1

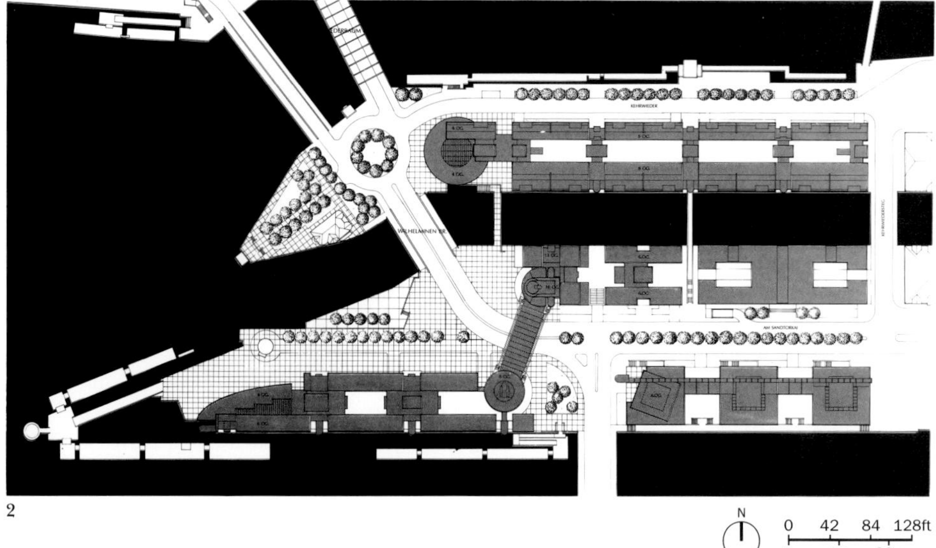

2

1 Sectional elevation from the west
2 Site plan
3 Model: view of the proposal in the context of Hamburg
4 Model: aerial view from the west

3

4

570 Lexington Avenue, The General Electric Building

Design/Completion 1988/1989
New York, New York, USA
The General Electric Company
Steel frame
Brick, terra cotta, aluminum-frame windows

The restoration of the exterior of the historic General Electric Building focuses on the replacement or preservation of historic fenestration, and the cleaning and replacement of brickwork and terra cotta details in accordance with the guidelines established by the New York City Landmarks Preservation Commission.

The renovation design specified the fabrication of new custom-designed aluminum windows that replicated the original steel-frame windows; the replacement of deteriorated terra cotta spandrels and other ornamental details with precast concrete that matches the color and texture of the undamaged terra cotta; the renovation of the lobby with a new aluminum-leaf ceiling; the renovation of street-level storefronts including marble refinishing; the restoration of the original steel-frame structure; and the reconstruction of the intricate, deteriorated metalwork that forms the building's crown. A dramatic lighting of the tower top restored the building to its rightful place among the skyscrapers on Manhattan's skyline.

1

2

3

4

1 General view of tower
2 Detail: tower top at night
3 Detail: tower top in daylight
4 Detail of restored facade

Southeast Financial Center

Design/Completion 1980/1985
Miami, Florida, USA
Gerald Hines Interests and Southeast Bank
1,800,000 square feet
Tube frame
Granite, marble, glass and metal

The 55-story office tower, faceted to provide multiple water views, contains a lower-level banking floor and structured parking for 1,148 cars. Between these two solid volumes is an open-sided breezeway planted with palms. A consistent play of matte and reflective surfaces characterizes the project, along with the widespread use of pale marble flooring, simple woods, and a palette of clear colors. The 178-seat auditorium on the fourth floor is walled with laminate and back-lit lacquer screens with roll-down blackout screens for film screenings. The acoustically responsive ceiling is of undulating rough plaster.

1

1 View from across Biscayne Bay
2 Covered plaza between office tower and parking structure

2

Citicorp at Court Square

Design/Completion 1987/1989
Long Island City, New York, USA
Citicorp at Court Square
1,400,000 square feet
Steel
Glass and metal curtain wall

On a direct axis with the midtown Manhattan Citicorp headquarters building, the building is placed on a two-acre site bounded by wide streets and small parks, a landmarked residential block, and a historic courthouse. The 50-story tower, with its curtain wall of glass spandrels enclosing pale green metal panels, features graduated corner setbacks at the highest floors, and is a defining element of the Long Island City skyline. A low-rise building adjacent to the tower houses ancillary services and retail space, while a seven-story glass-enclosed rotunda is the focal point for the entrance plaza, two subway stations, and the pedestrian concourse entrance.

The design for this "intelligent building" incorporates computerized building management systems for life safety, security, HVAC, and energy savings. Easy-access raised flooring throughout all office space permits the rapid installation and renovation of power, telephone, and computer signal needs.

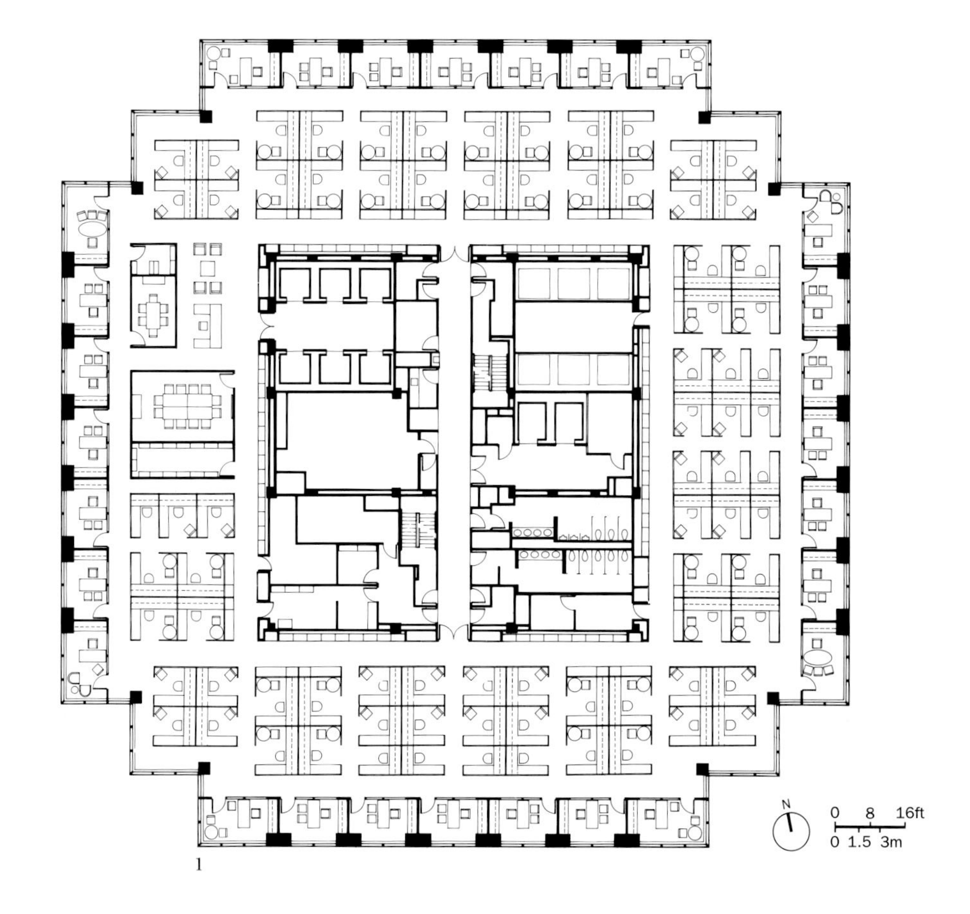

1

2

1 Typical floor plan
2 Detail of cladding
3 View from Manhattan

3

Dearborn Tower

Design/Completion 1988
Chicago, Illinois, USA
Bramalea, USA
2,500,000 square feet
Reinforced concrete core with structural steel superstructure
Granite, stainless steel, painted aluminum

The project is located within a concentration of government and commercial offices along the Dearborn Street corridor. The five-story base occupies the entire site, and is related to similarly scaled commercial buildings as well as the surrounding retail environment. The base is part of a continuous sequence completing the wall of masonry facades that surround the Federal Center. As the Federal Center buildings are freestanding steel-and-glass objects within a masonry wall, the tower is conceived as a glass-and-steel object emerging from within the enclosure.

The structural system incorporates a central core element of concrete shear walls in a cruciform configuration. The exterior structure is tied to the core with two-story-deep trusses at the mechanical floor levels, expressed as buttresses at the crown. A glass and stainless steel curtain wall provides unobstructed views from all office floors, angling away from the central figure in a series of sawtooth-like setbacks.

1

1 Model: tower top
2 Model: Dearborn Street elevation

2

AT&T Corporate Center

Design/Completion 1986/1989
Chicago, Illinois, USA
Stein & Company
1,700,000 square feet
Steel frame
Polished granite, gray glass, painted aluminum and bronze

The 60-story tower, located at the corner of Monroe and Franklin Streets, is an example of the Chicago skyscraper, as exemplified by the Chicago Board of Trade, the Carbide and Carbon Building, and Eliel Saarinen's runner-up entry to the 1922 Chicago Tribune Tower Competition.

In the style of the massive buildings of the surrounding business district, the tower rests on a highly articulated, five-story granite base with a monumental entrance on Monroe Street. At the 29th and 44th floors, five-foot setbacks taper the mass of the building. The building changes in color from deep red at the base, to lighter red at the lower floors, to a light rose color for most of the shaft. Recessed spandrels and details are a deep green granite with a decorative abstract pattern.

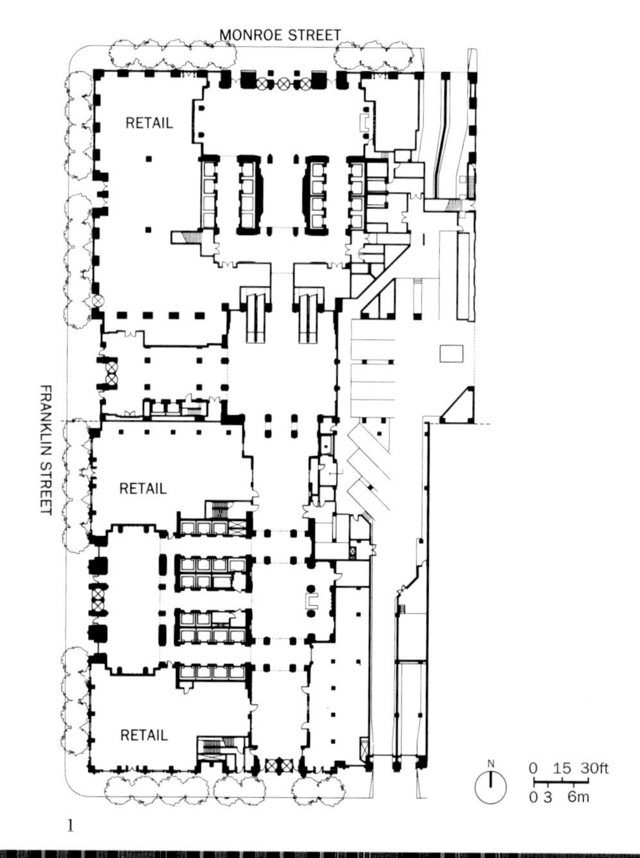

1

2

1 Ground-floor plan of phases 1 and 2
2 View of entry
3 View of tower

3

Fubon Banking Center

Design/Completion 1990/1994
Taipei, Taiwan
ARTECH Inc.
580,000 square feet
Steel frame
Granite, aluminum, glass

The Fubon Banking Center fronts on Jen Ai Road, a main artery in Taipei, and commands views of the mountains to the north. The 25-story tower, clad in granite, aluminum, and glass, is crowned by a two-story space that houses a boardroom and special events rooms. This space is enclosed by an angled glass plane etched with the bank's logo, acting as a billboard for the city's commercial district. At the ground level, skylights that are reminiscent in form of Chinese lanterns reinforce the street edge and define an open plaza. High-performance glazing dominates the building skin because of its durability in the harsh Taiwanese climate.

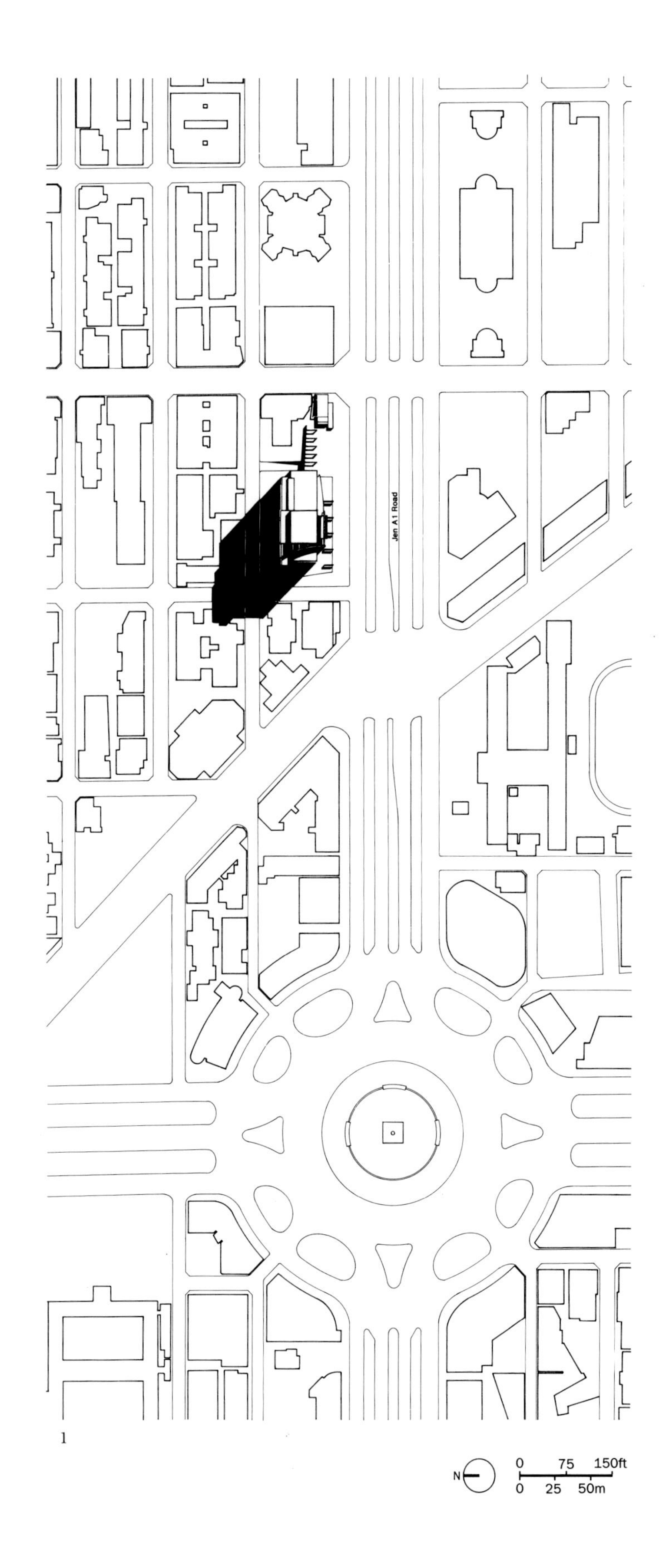

1 Site plan
2 View of south-east corner

2

National Commercial Bank of Jeddah

Design/Completion 1979/1983
Jeddah, Saudi Arabia
National Real Estate Company of Jeddah
766,000 square feet
Concrete
Precast concrete faced with travertine

The site for this project is a three-acre plaza on the edge of the Red Sea. The project's geometry consists of a triangular 27-story office tower juxtaposed with a six-story circular garage. The configuration of the site and local climatic conditions generated the overall form of the complex.

The verticality of the bank tower is interrupted by three dramatic triangular courtyards chiseled into the building's facade. Two of these courtyards, each seven stories high, face south towards the old portion of the city and the Red Sea. The third nine-story courtyard faces north-west towards the sea. Office windows open directly onto these courtyards with an inward orientation typical of Islamic traditional design.

Interior spaces feature exquisite materials and finishes including black granite and marble. Dining facilities include a luxurious executive dining floor and a colorful employee cafeteria.

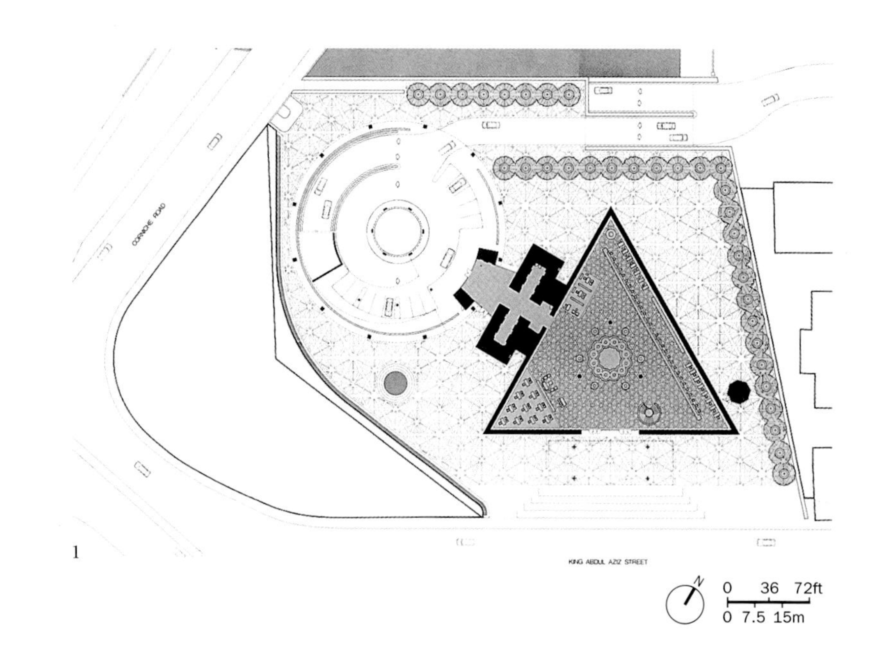

1

2

1 Site plan
2 The garage structure from above
3 General view

3

4

5

4 Banking hall
5 View of Jeddah from the office tower

Plaza Rakyat

Design/Completion 1993/1998
Kuala Lumpur, Malaysia
Daewoo Corporation
6,600,000 square feet
Reinforced concrete
Stainless steel, glass, metal panels, exposed concrete

The mixed-use development and transportation center was conceived as a group of buildings centered around a rooftop civic garden. The ensemble is anchored by a 77-story office tower that punctuates the southern edge of the triangular site. Together with the office tower, a continuously arcaded, multi-level shopping center creates a vibrant civic facade along the site's most active boundary. At the far corner of the retail block is a 415-room, four-star hotel linked to a 315-unit serviced apartment complex along the residential edge of the site. Future expansion has been planned to include an exhibition/conference center linking the project to the adjacent tourist district.

The design addresses the triangular site with different facades that respond to the specific scales of surrounding streetscapes. Each component has a separate identity but utilizes common exterior wall materials and detail palette.

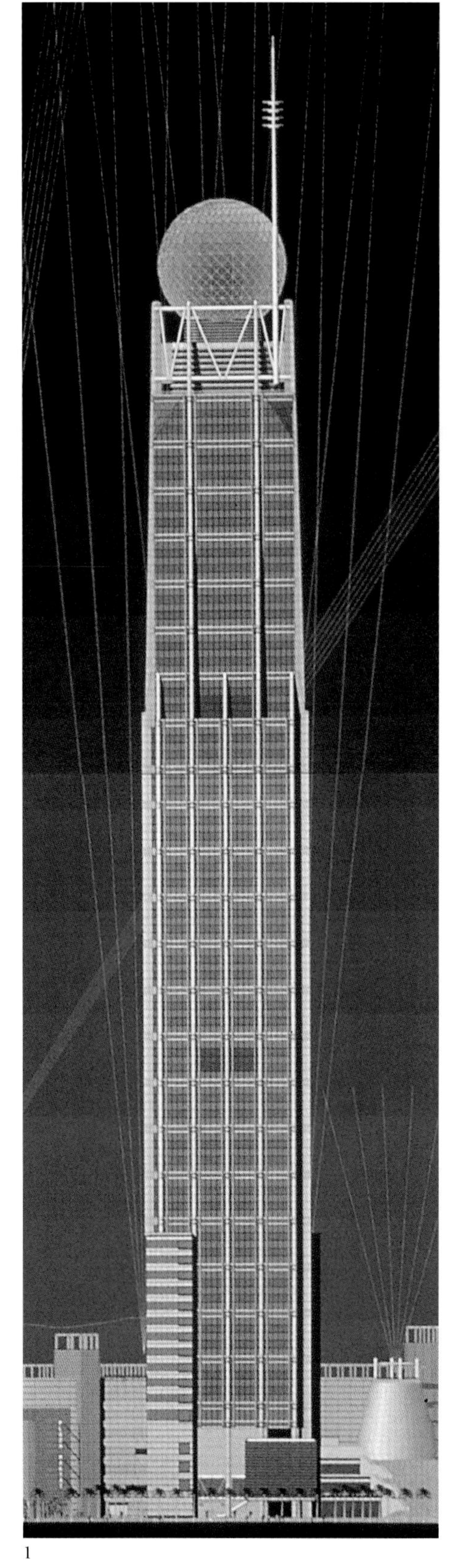

1

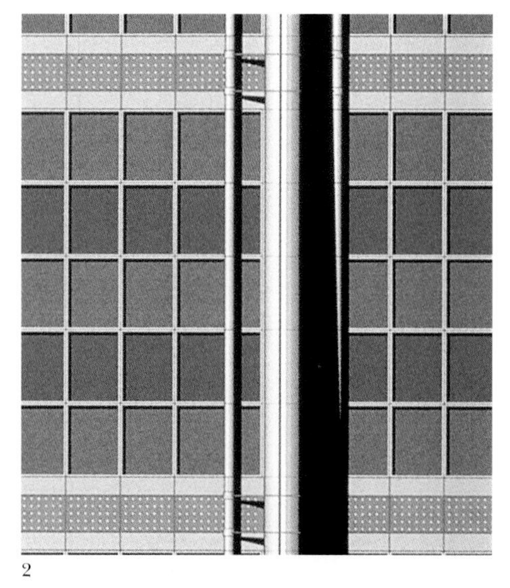

2

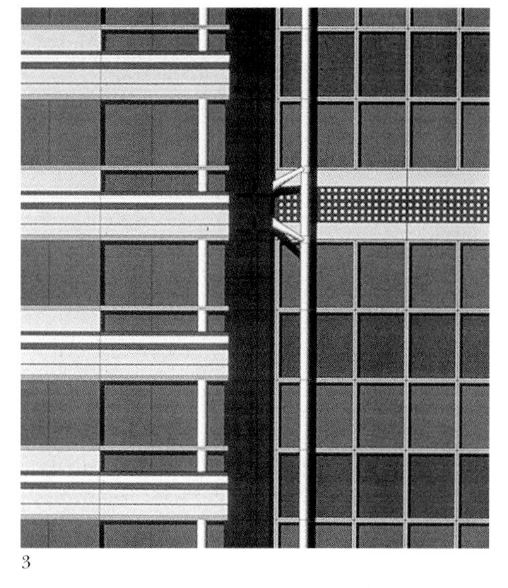

3

4

1 Office tower: street elevation
2 Office tower: upper wall
3 Office tower: lower wall
4 Retail arcade
5 Project overview looking west

5

Gas Company Tower

Design/Completion 1988/1991
Los Angeles, California, USA
Maguire Thomas Partners
1,780,000 square feet
Steel frame
Granite, aluminum panels, reflective glass and metal curtain wall

Linking two of the city's public precincts, the 55-story tower updates the Southern California Gas Company's image. Rising from the blue/gray granite shaft is an elliptical blue glass volume; its shape symbolizes a gas flame and its reflectivity distinguishes it from the building shaft. The design reduces in mass towards the top, responding to the downtown site with different building skins. The central shaft is simple in form like the neighboring Arco Tower, with a granite facade featuring punched openings. The side volumes are wrapped in metallic skins echoing an adjacent building.

The base of the tower is a critical link between the Central Library and Pershing Square, both newly renovated. The Pershing Square entrance is proportioned to relate to the openness of the park and to the historic Biltmore Hotel, with the corner of the four-story base cut at the height of the hotel.

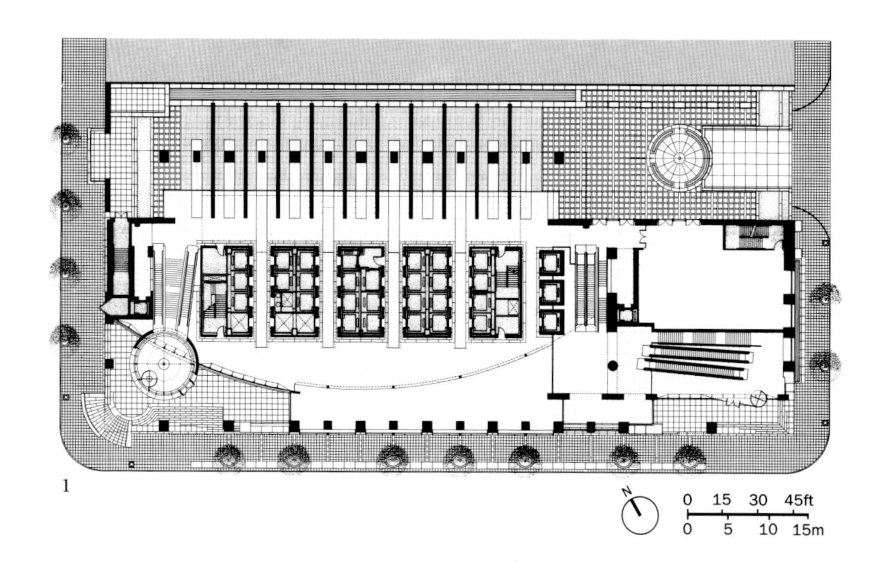

1

2

1 Lobby-floor plan
2 View of an elevator lobby
3 Tower in the skyline

3

388 Market Street

Design/Completion 1982/1986
San Francisco, California, USA
Honorway Investment Corporation
500,000 square feet
Structural steel
Polished red granite, clear glazing, copper dome

The plan of this mixed-use tower, generated by its site at the 36-degree intersection of two street grids, echoes Market Street's historic flatiron forms. The plan symbolizes the building's pivotal position in the financial district, while also referring to the cylindrical 101 California Street building. The program comprises 16 floors of offices, six floors of residences, retail facilities, a health club, and parking.

The first two stories form a 40-foot high retail base that aligns with three historic buildings facing the site. A two-story public gallery bisects the base to connect Pine and Market Streets, serving separate lobbies for the office tower, apartments, retail, and garage.

Six full floors of apartments are situated at the top of the building, with 70 per cent of the perimeter open to the outside. The top floor of apartments and mechanical space provides a strong cornice line. A copper dome encloses the cooling towers, echoing the cylindrical building next door.

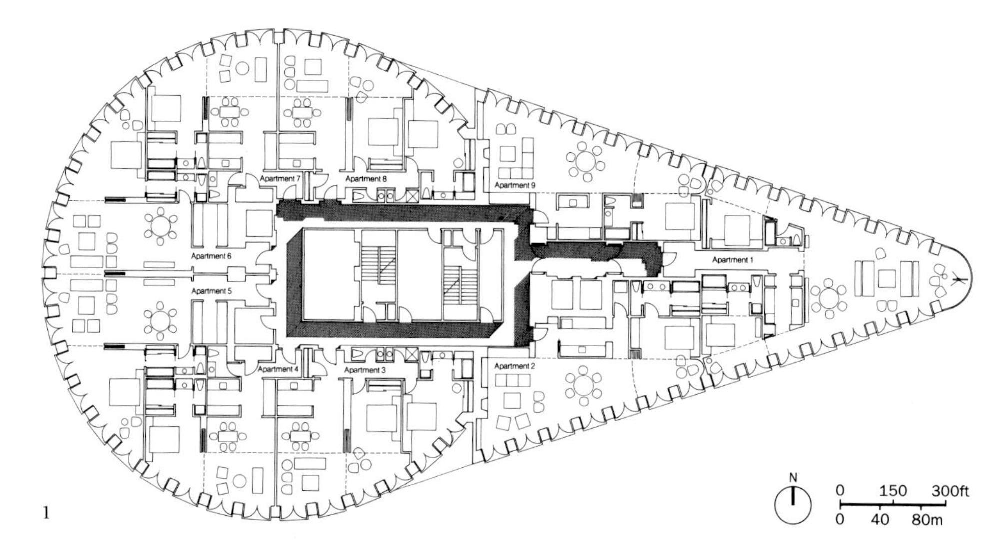

1

2

1 Ground-level plan
2 West side view from Market Street
3 East side view from the corner of Market and Front Streets

3

303 West Madison

Design/Completion 1984/1988
Chicago, Illinois, USA
Jaymont Properties Inc.
350,000 square feet
Steel frame
Granite, glass block, metal details

The tower design reinterprets basic elements that characterized the Chicago School: distinct base, middle, and crown elements; tripartite oriel windows; and ground-floor retail. The lower floors define the base of the building, while the third floor makes a transition between base and shaft. At the building top, a single story between the shaft and a two-story loggia with open roof coffers defines the crown while reiterating the structural grid.

The wall surface is articulated with oriel windows and layers of light gray granite with polished green granite trim. White-painted window frames and mullions with horizontal red accent stripes on the two-story spandrels characterize the fenestration. Reflective glass "buttons" at the center of each spandrel refer to the adjacent reflective glass tower.

1

2

1 View of Franklin Street facade with open roof coffers
2 Franklin Street entry with stained glass windows
3 Oblique view along Franklin Street

3

100 East Pratt Street

Design/Completion 1991/1992
Baltimore, Maryland
IBM Corporation
420,000 square feet
Steel and concrete
Stone, precast concrete, metal panel curtain wall

The architectural vocabulary of 100 East Pratt Street is an extension of the modernist context that has evolved along Pratt Street and around the Inner Harbor. Set back from Pratt Street, the envelope maintains the 10-story cornice line of its neighbors, with the 60-foot wide, 28-story tower connected to the north side of IBM's existing building. The north facade, facing the downtown area, is realized in stone, masonry, and glass, while the south side features a glass and steel bay window overlooking the harbor.

The tower is crowned by a series of exposed steel trusses that support the projecting south facade. In its delicacy of tensile expression, the truss series evokes the maritime imagery of the waterfront. The trusses also provide the wind bracing required by the very narrow tower profile that allows its northern neighbors uninterrupted views of the Inner Harbor.

1

1 Lobby interior
2 View of tower from the harbor

2

United Gulf Bank

Design/Completion 1982/1987
Manama, Bahrain
United Gulf Bank
98,000 square feet
Steel frame
Precast concrete, glass

This 12-story office building located in the Diplomatic Quarter is responsive to the physical, climatological and cultural conditions of its context. The curved facade mediates between the public realm of the street and the private realm of the bank. A three-story arcade at ground level shields pedestrians from the harsh sun, dust, and noise of the street, echoing the regional vocabulary of arcades and shaded streets.

Windows are deeply recessed in the exterior wall. Vertical green glass fins, oriented north–south, effectively intercept lower east and west sunlight, evoking the *mashrebeeyah*, the indigenous latticed sunscreen. Open office areas surround a 10-story atrium, bridged by glass block terraces that allow daylight to penetrate into all interior spaces. External light is admitted to the atrium through a pattern of narrow slots in the thick north-east wall.

1

1 View from the east
2 Section through the atrium
3 Level 10 plan
4 Tapestry of United Gulf Bank

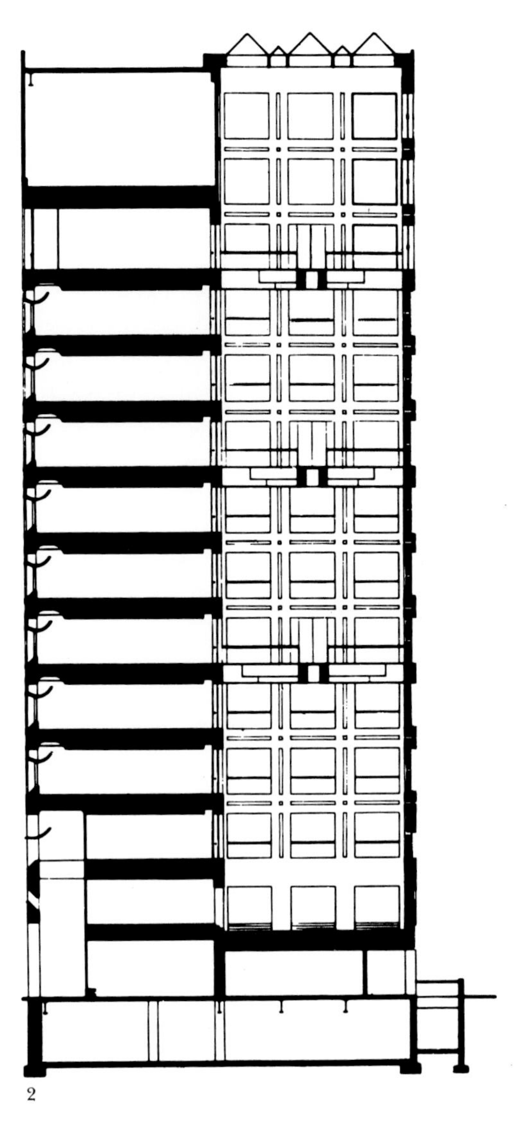

2

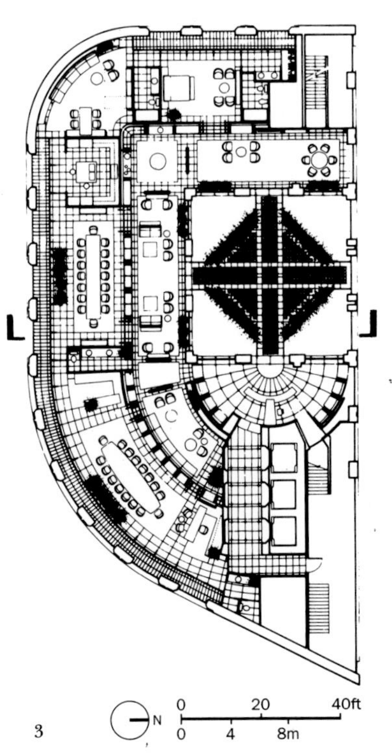

3

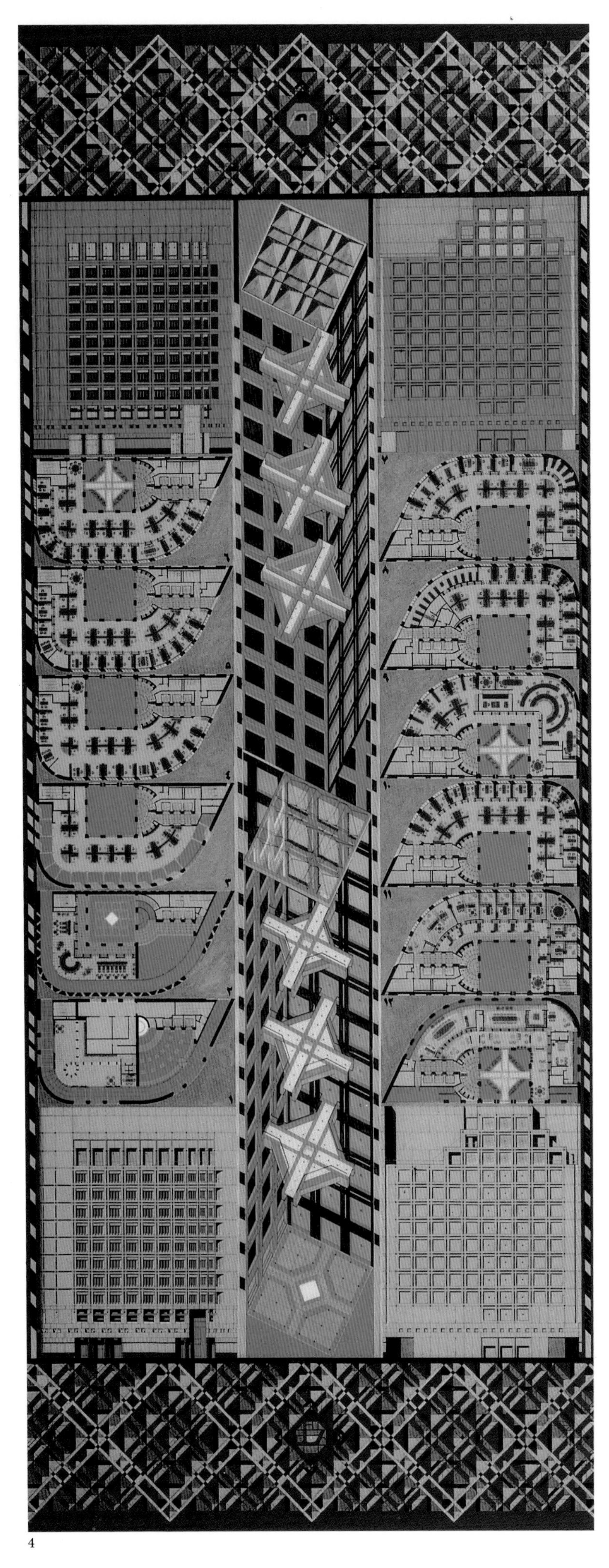

4

5 Detail of a typical wall
6 Atrium at the executive level

5

6

7 Atrium view from level 2
8 Bridge view at night

7

8

Birmann 21

Design/Completion 1992/1996
Sao Paulo, Brazil
Birmann S.A. Comercio e Empreendimentos
375,000 square feet
Concrete structure
Glass and metal curtain wall, Brazilian granites, marble

Designed to be the regional headquarters for international corporations in Sao Paulo's newest commercial sector, the 24-story office tower includes an executive dining club, a conference center, a cafeteria, a health club, and public gardens.

The design is a harmonious composition of vertical forms varying in texture, materials and color, the focus of which is an external steel skeleton that culminates in a 490-foot spire. The spire and the building's placement perpendicular to the city's peripheral highway establish a presence on the skyline.

The major glazed facade is curved slightly towards the south for the proper orientation in the southern hemisphere, and provides views of the Pinheiros River. The northern facade, including the main entrance, is clad in Brazilian granites and marble, and features smaller, deeper windows to reduce cooling loads.

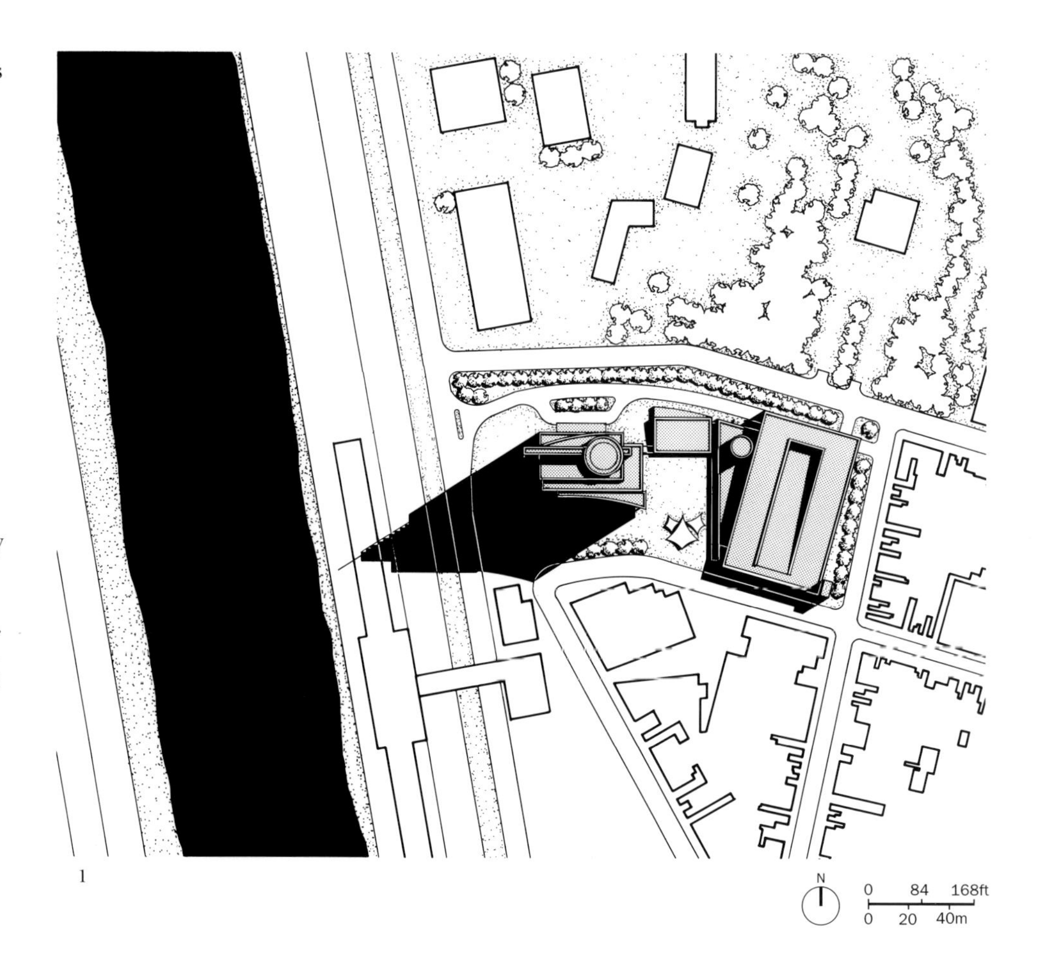

1

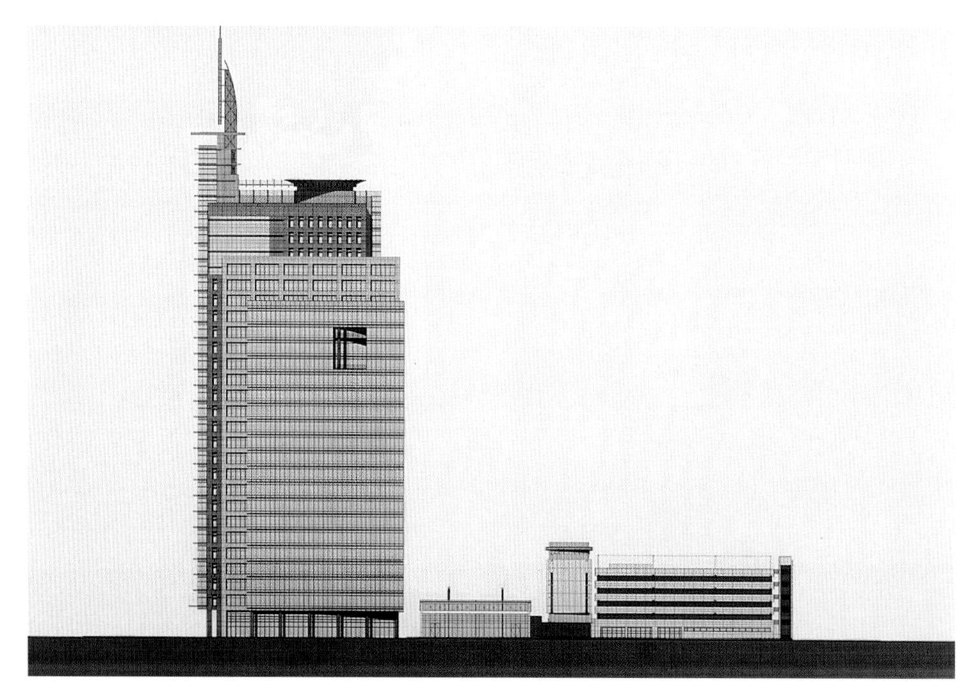

2

1 Site plan
2 Computer rendering of south elevation
3 Model: south elevation

3

The World Center

Design/Completion 1992/1995
Manila, The Philippines
MegaWorld Properties & Holdings Inc.
34,000 square meters
Concrete frame
Glass and metal curtain wall

Located in the Salcedo/Legaspi area in Makati, the core of Manila's financial district, the 30-story tower is clad in metal and blue-green glass. The design scheme is unusual for the surrounding "party wall" context in that it places the tower adjacent to a low-rise element, allowing for three fenestrated facades rather than two. The curtain wall design varies from facade to facade in response to climate and solar considerations, featuring brises-soleil on the south, and a slight angling of the east facade.

Built to the maximum allowable height, the project features a 500-foot spire that identifies the building on the Manila skyline and functions as a telecommunications tower. The enclosed balconies punctuating the upper floors of the curtain wall indicate the executive club areas. The three-story component incorporates a two-story lobby entrance and retail.

1

1 Entry perspective
2 Model

2

Aramco

Design/Completion 1993
Dhahran, Saudi Arabia
350,000 square feet
Steel frame
Glass, stainless steel, granite

The complex is planned in three phases, and incorporates new office and mixed-use buildings within an existing campus of technical, research, and administrative buildings. The focal point of the complex is a new 20-story corporate headquarters tower consisting of two curvilinear pieces of 10 and 20 stories that enclose a 10-story central atrium.

The tower's exterior wall is a collection of expressed components, materials, and shapes. The effects of the harsh desert sun are controlled by brises-soleil of vertical glass fins, stainless steel trusses, and horizontal sunshades. The broad eastern facade is broken by a set of stacked three-story atria set against a 12-story black granite shear wall that ties the soft curving forms to the rectilinear geometry of the existing buildings.

A covered two-story galleria terminates in the atrium of the new corporate building and features a 200-foot-long exhibition wall.

1 View from the south
2 View from the west

1

2

Chase Financial Services Center at MetroTech

Design/Completion 1991/1993
Brooklyn, New York, USA
Chase Manhattan Bank
1,860,000 square feet
Steel
Brick veneer, stone

This two-building complex responds to the master plan and design guidelines for a new, high-technology urban office precinct, and is the major operations center for the bank.

The design combines simple geometric forms with highly articulated surfaces throughout. A solid stone band at the base enhances the pedestrian experience, while glass and metal storefronts and evenly spaced piers relieve the long building wall and lead into the arcade along the north facade. Corbeled brick corners and other details echo the scale and character of the adjacent landmarked 1830s houses.

The larger lower floors serve as a podium for the slender 23-story tower, whose verticality is emphasized by articulated corners. At the top, the vertical metal fins that conceal the cooling towers and functioning air intakes are backlit to create a building crown that distinguishes the bank on the Brooklyn skyline.

1

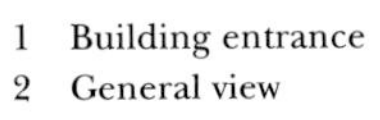

1 Building entrance
2 General view
3 Lower lobby

2

3

4

4 Upper lobby
5–6 Elevator lobby

5

6

Spiegel Corporation

Design/Completion 1989/1992
Downers Grove, Illinois, USA
Hamilton Partners Inc.
660,000 square feet, parking for 2,100 cars
Steel frame
Precast concrete, granite panels, aluminum and green glass curtain wall, ceramic-coated glass spandrels

The development is set in a rolling landscape adjacent to a forest preserve. The first phase of a three-phase project, the 13-story office building is designed to present two separate images. The formal east facade is clad in precast concrete and light gray granite panels with an aluminum and green glass curtain wall; the west side is characterized by a gently curving curtain wall of green tinted glass and bands of green ceramic-coated glass spandrels.

Typical office floors are 45,000 square feet each, with a sample room on each of the first two levels. The project incorporates executive offices and boardrooms on the top level. Other components include a conference center; a corporate reception area with a 36-screen video display; a four-story atrium containing an indoor garden; employee dining in a separate one-story pavilion connected to the main building; and a six-level parking garage.

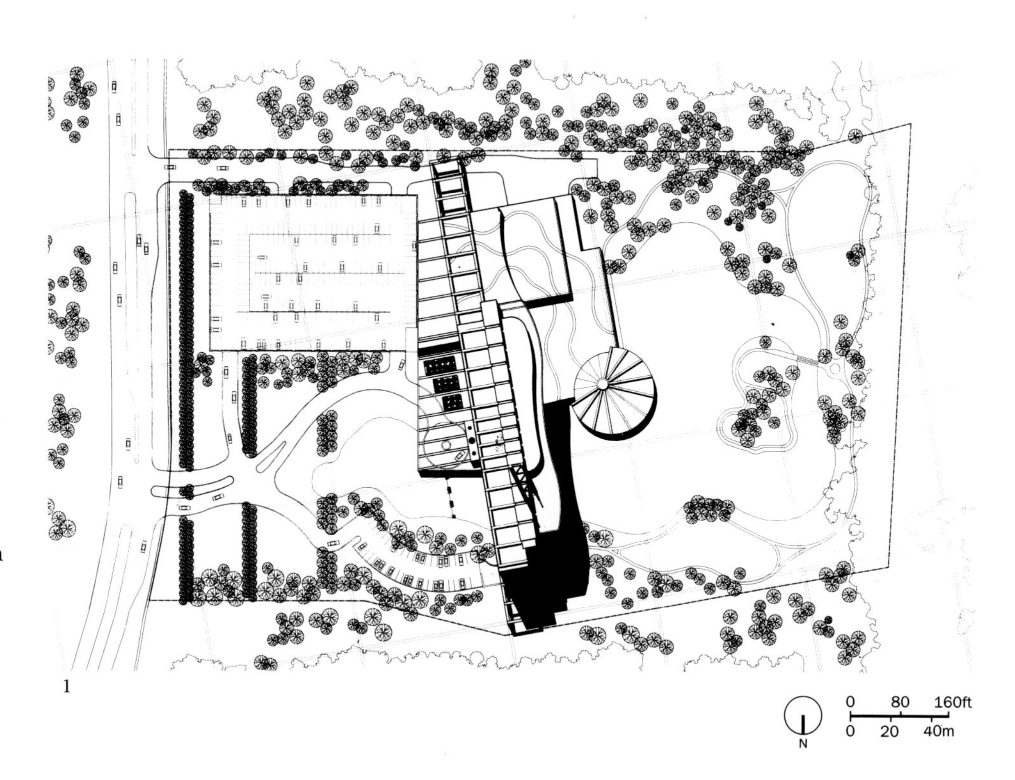

1

2

1 Site plan
2 View from the east
3 Curvilinear facade facing the forest preserve to the west

3

Das American Business Center at Checkpoint Charlie

Design/Completion 1992/1996
Berlin, Germany
Checkpoint Charlie Bauprojekt GmbH
446,700 square feet
Concrete
Glass and metal curtain wall, limestone

Block 105, part of the American Business Center at Checkpoint Charlie, marks the intersection of the demolished Berlin Wall along Zimmerstrasse, and the pre-Baroque wall of the old city of Berlin.

Designed for residential, retail, and office use, the complex incorporates an existing building into three new structures, and has been conceived so that the components can function individually or as one unit. The first component is a contemporary interpretation of the historically prevalent streetwall building that fully or partially encloses internal courtyards. At the intersection of Friedrichstrasse and Zimmerstrasse, the bar-shaped building flows into a rounded corner that becomes a circular tower, highlighting the importance of the site. This architectural gesture focuses the complex on the former Checkpoint Charlie gate.

The courtyard building is the third component, a U-shaped structure clad in glass and limestone with painted metal details, completely enclosing a central public court.

1

2

1 Zimmerstrasse elevation
2 Aerial view
3 Model of tower

3

Industrial and Commercial Bank of China

Design/Completion 1993/1996
Beijing, China
Industrial and Commercial Bank of China
1,000,000 square feet
Steel moment frame
Metal and glass

Located on Chang An Street, the ceremonial entry to Tiananmen Square, the siting of the bank reflects fundamental principles of Chinese urban design. The plan organizes the building axially, incorporating geometric forms that refer to historic precedents. The 150-foot height limit, 100- to 115-foot setbacks, and a double row of trees along the boulevard enhance the civic and monumental presence of the project.

The design for the new headquarters provides a hierarchy of major spaces that are designed for future expansion, and incorporates recreation areas, an interior garden plaza, and parking for automobiles and bicycles. The south-facing ceremonial entry, at the main facade, is approached by a controlled-access driveway for the use of arriving dignitaries. Taxi drop-off points and access to surface and below-grade parking are provided at east and west employee entries. The careful massing results in only minimal shading of the adjacent buildings.

1

1 Interior detail of curved wall
2 View of main entrance from south corner
3 Aerial view

2

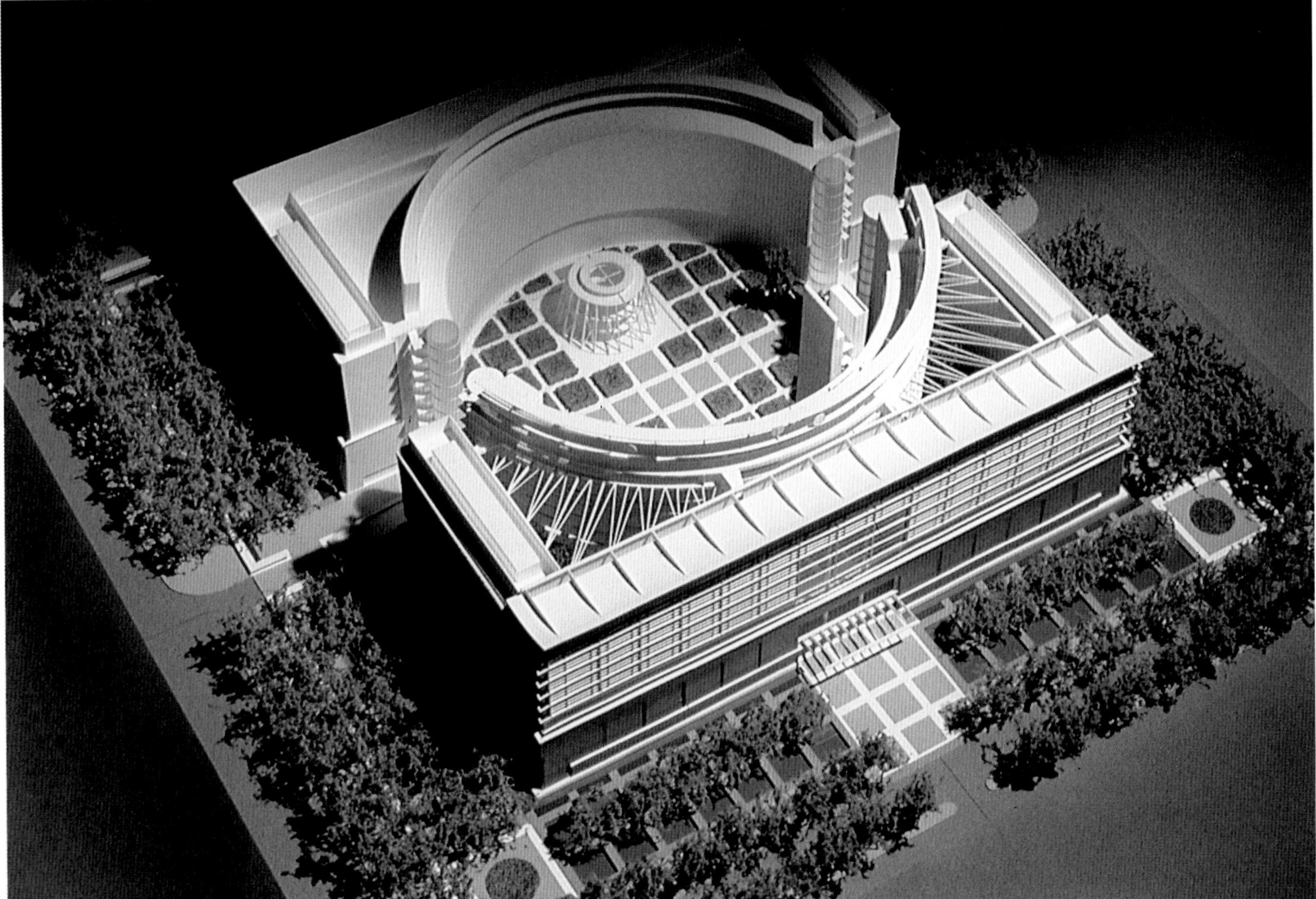

3

Centro Bayer Portello

Competition 1991
Milan, Italy
Bayer Milan
Precast concrete frame
Clear and fritted glass and stainless window wall, marble sunshades, marble base cladding, stainless steel roof

A major corporate presence for Bayer in Milan generates this competition scheme. The main entrance is in the center of the whole complex, with the building's mass to the north and south sides of the street connected by bridges above street level. Over the main entrance is a great illuminated glass and steel crown containing the symbols of Bayer, Agfa-Gaevert, Miles Laboratories and other corporations housed within the complex. The crown is clearly visible from the autostrada and all approaches to the building, and is immediately recognizable.

The design solution creates an internal galleria, rising the full height of the building, which is the heart and communications center of the complex. In form it is an elongated triangle, reflecting the street pattern around the site. In function it is both the *piazza centrale* of the Bayer community and also the organizing element from which all components of the complex are visible and easily reached.

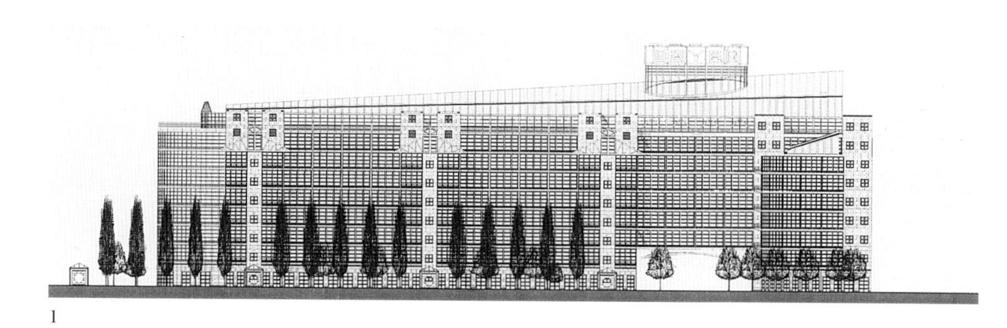

1

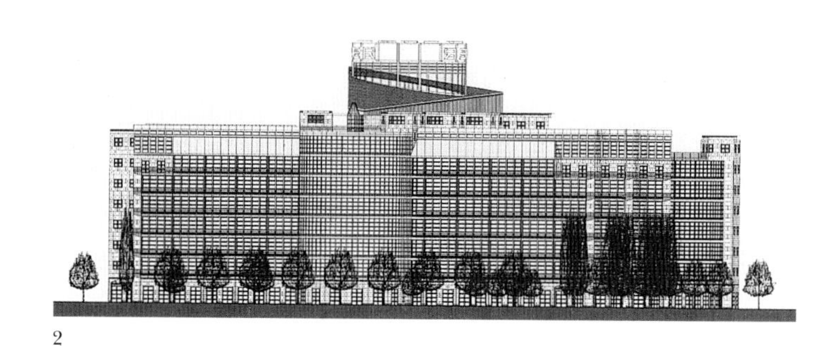

2

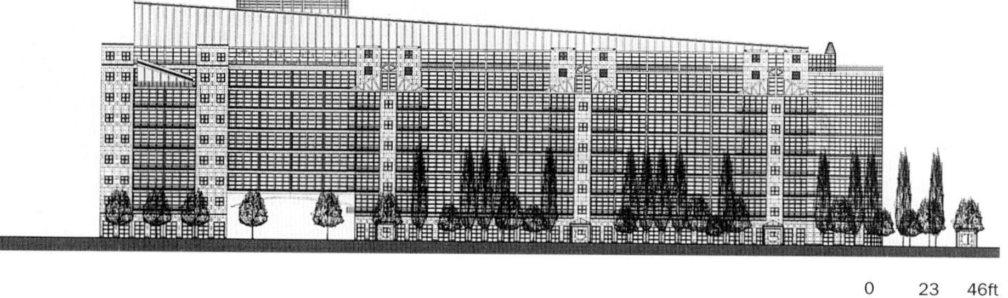

3

1 East elevation
2 South elevation
3 West elevation
4 Roof plan
5 Aerial view and detail of entry gateway

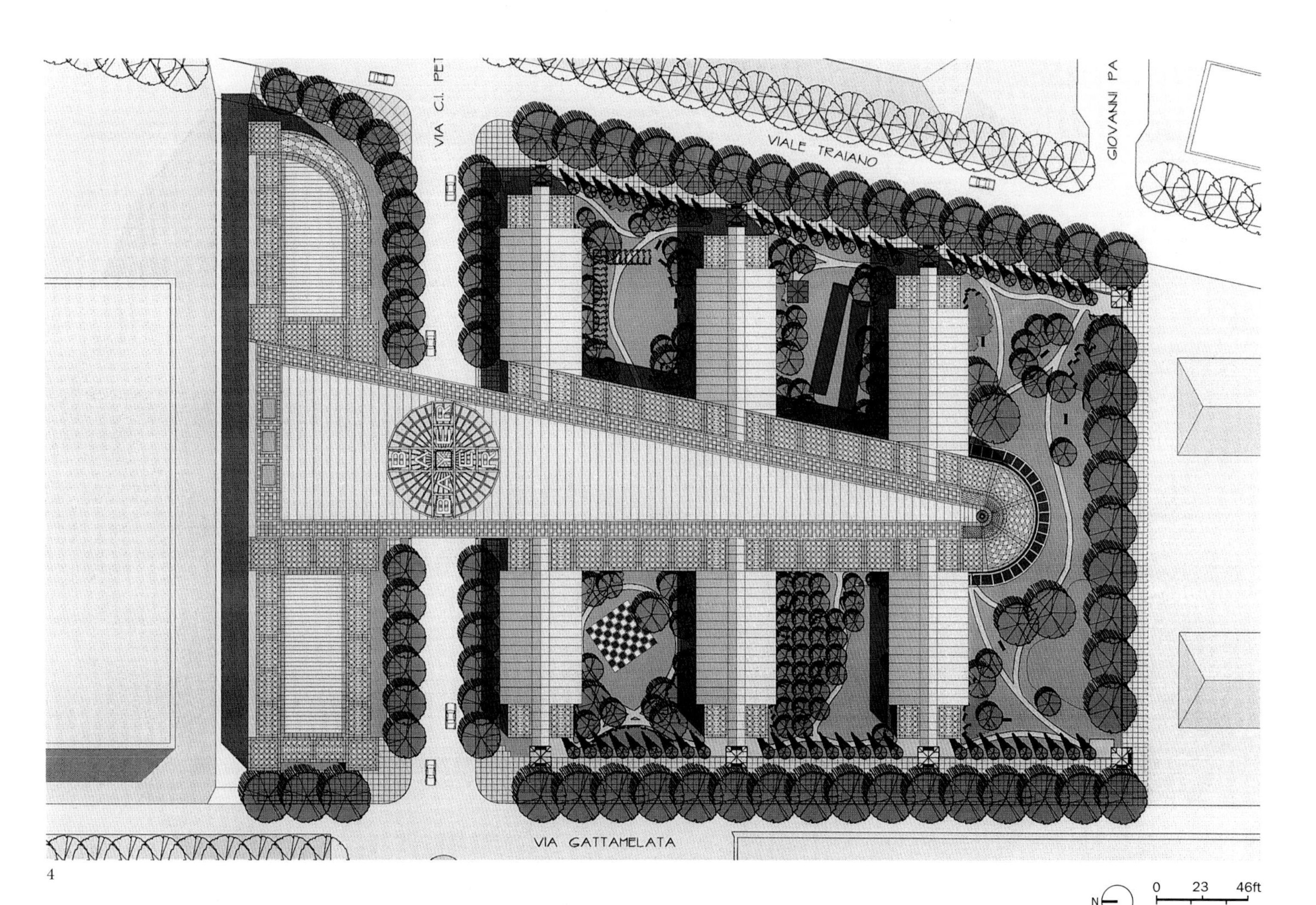
VIA C.I. PET
VIALE TRAIANO
GIOVANNI PA
VIA GATTAMELATA

4

N
0 23 46ft
0 5 10m

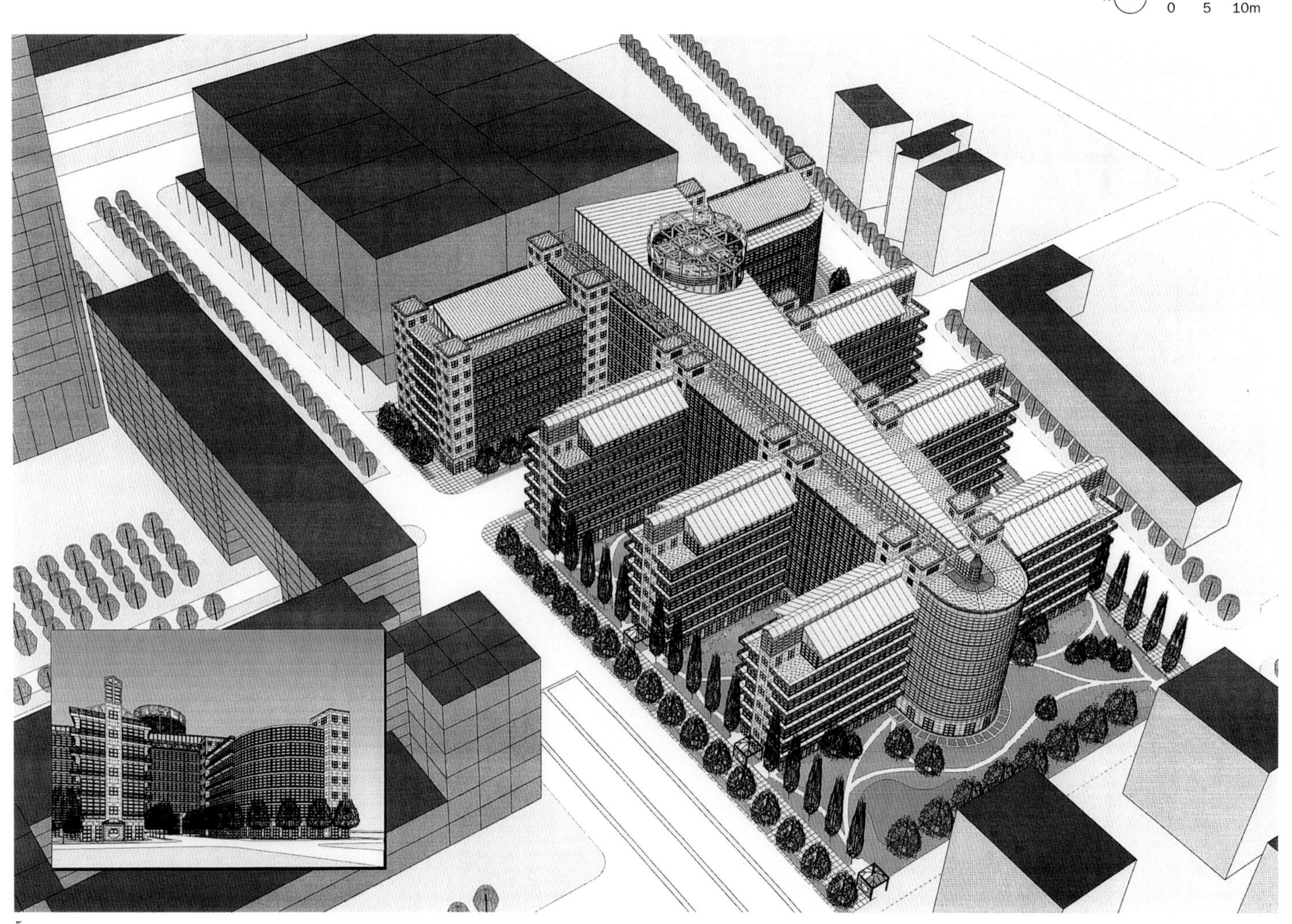

5

Hotel de las Artes at Vila Olimpica

Design/Completion 1990/1992
Barcelona, Spain
The Travelstead Group
1,176,118 square feet (109,000 square meters)
Exposed steel superstructure
Exposed steel high-rise building with stone base; concrete office building; low-rise concrete, steel, and wood beachfront development

The multi-use complex, with a five-star hotel, apartment residences, a village-like retail center, and 128,000 square feet of first-class office facilities, provides a node of activity within a defined waterfront environment. Located on the Mediterranean coast close to the Ramblas, the complex was the result of a bold vision to bring this part of Barcelona to the sea.

The complex combines the amenities of an international business hotel with outstanding resort facilities. Restaurants and lounges, ballroom facilities, health clubs and a swimming pool afford travelers the opportunity to mix business events with leisure activities and exercise. In addition to the 456-room business hotel, the hotel tower offers 29 apartment residences.

Located directly across Francisco de Aranda from the hotel and retail complex, the five-story office building encircles a raised plaza and is ideally suited for meetings. Open floor plans and flexible planning characterize the interior program.

1

2

3

1 Retail and office building
2 Hotel port cochere
3 Seafront retail area
4 Elevation from Mediterranean Sea

4

5

5 Exposed steel frame at lower base
6 Hotel entry at lower base
7 Exterior garden detail
8 Exterior garden at retail level

6

7

8

Naval Systems Commands Consolidation

Design/Completion 1992/
Arlington, Virginia, USA
General Services Administration
1,400,000 gross square feet
Structural steel frame, composite metal deck
Glass and precast concrete; poured-in-place concrete for parking structure

Located one mile from the Pentagon, the building will consolidate the various departments of the Naval Systems Commands. The need for flexibility in plan for the various commands over the next 100 years suggested a footprint with large areas of uninterrupted space.

Situated on the eastern half of the 6-acre site, the project's mass is a linear configuration composed of two bars of seven and nine stories, separated by common core elements and atria. The central portion of the site is perceived as the front lawn of the building, leading to the formal main entrance. To mitigate the impact of the project on the surrounding community, the entire western half of the site was conceived as landscaped open area.

1 View of central atrium
2 Model: aerial view from south-west
3 Model: aerial view from north-east

1

2

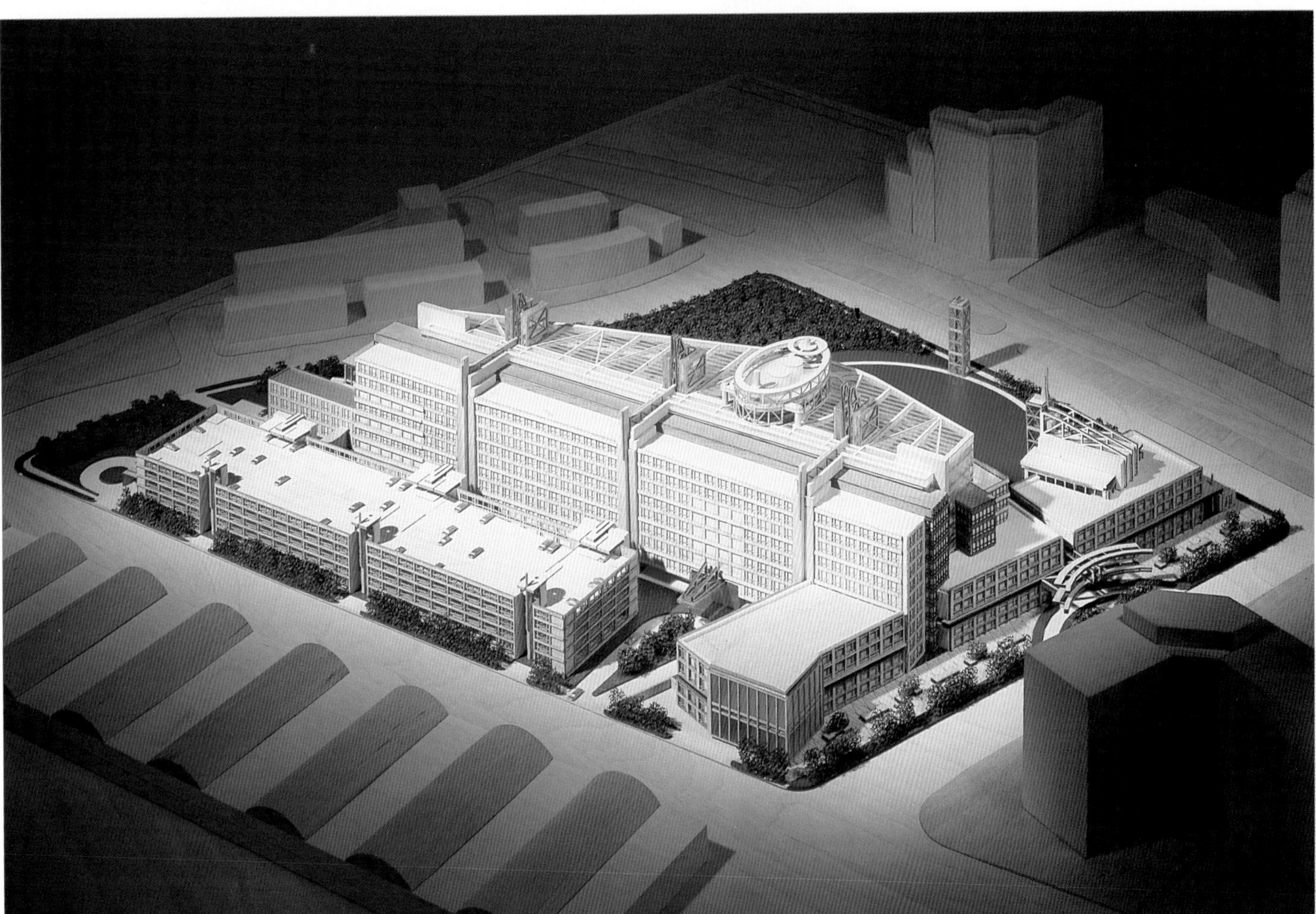
3

Administrative Centre for Sun Life Assurance Society

Design/Completion 1991/1996
Bristol, Avon, England
Sun Life Assurance Society PLC
550,000 square feet
Concrete frame
Brick, reconstituted stone panel, glass block, metal panel

Situated adjacent to Parkway Railway Station approximately 20 miles from central Bristol, the site consists of three separate tracts of land, two of which are developed during Phase I, with the third to be activated during Phase II.

The project comprises about 550,000 square feet of space in two phases of 400,000 square feet and 150,000 square feet. Phase I operates as a new administrative center for the Sun Life Assurance Society. The company, which currently owns several leases in central Bristol, plans to relocate 2,400 employees to this location. Phase II is planned for future expansion and a maximum occupancy of 3,300.

Facilities within the complex include staff training centers, management suites, staff restaurant, retail facilities, and a sports center, as well as general office space. All functions within the building revolve around the internal "street" which links programmatic components and provides semi-public meeting places for building occupants.

1

1 Computer rendering of the internal street
2 Aerial view of model—night
3 Aerial view of model—day

2

3

Stockley Park

Design/Completion 1988/1990
Hillingdon, Middlesex, England
Stockley Park Consortium Ltd
320,000 square feet
Steel frame
Precast concrete

Set in an extensively landscaped office park, the three-building complex is located on an uneven triangular site bordered by a raised highway to the east and a canal to the south. The basic architectural plan adopted for all three buildings uses a 30-foot-square bay. This is logically expressed in an identical square plan for two of the buildings, while the third incorporates the same bay size but is irregular due to site constraints.

A white painted aluminum curtain wall, glazed with a combination of clear and fritted glass, is used for the two upper floors, with the ground floors clad in precast concrete. The exterior wall at the top floor is set back to expose the perimeter steel columns, and to emphasize the roof structure. Mechanical rooftop units, concealed behind colorful metal screens and skylights capping the central lightwells, result in an overall sculptural quality.

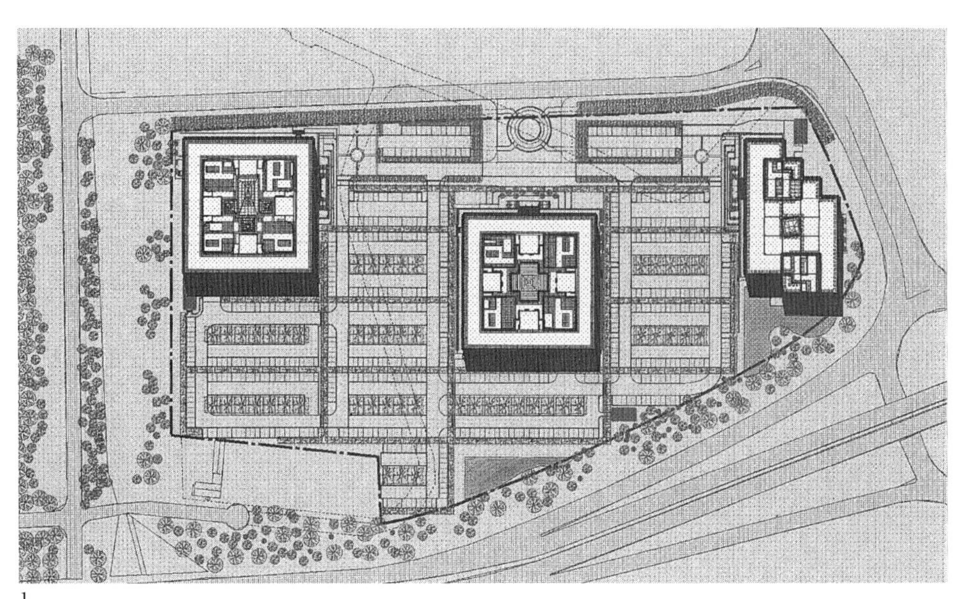

1

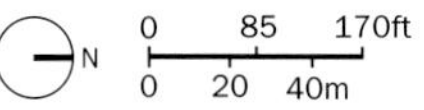

2

1 Site plan
2 Entrance to Building 9
3 Disabled access lift
4 Entrance to Building 9
5 External fire escape stair
6 Typical window wall detail

3

4

5

6

Chase Manhattan Bank Operations Center

Design/Completion 1991/1993
Yokohama, Japan
Chase Manhattan Bank
100,000 square feet
Steel frame
Glass and metal curtain wall, red Indian sandstone

Situated in the Kohoku New Town planned community, the site is organized around an enclosed garden precinct that provides space for cultural activities, and is adjacent to a gallery. A conceptualized relationship between building and landscape dominates the design.

In response to the topography of the site, the buildings have been developed on three levels, with office blocks on upper and lower levels, and an intermediate garden level.

The office buildings are simple geometric forms, featuring column-free areas that are glazed on three sides. Support functions and enclosed offices adjacent to the open office area are articulated as a vertical solid of red sandstone that forms a backdrop to transparent circulation and core elements. The single-level gallery is integrated into the garden design.

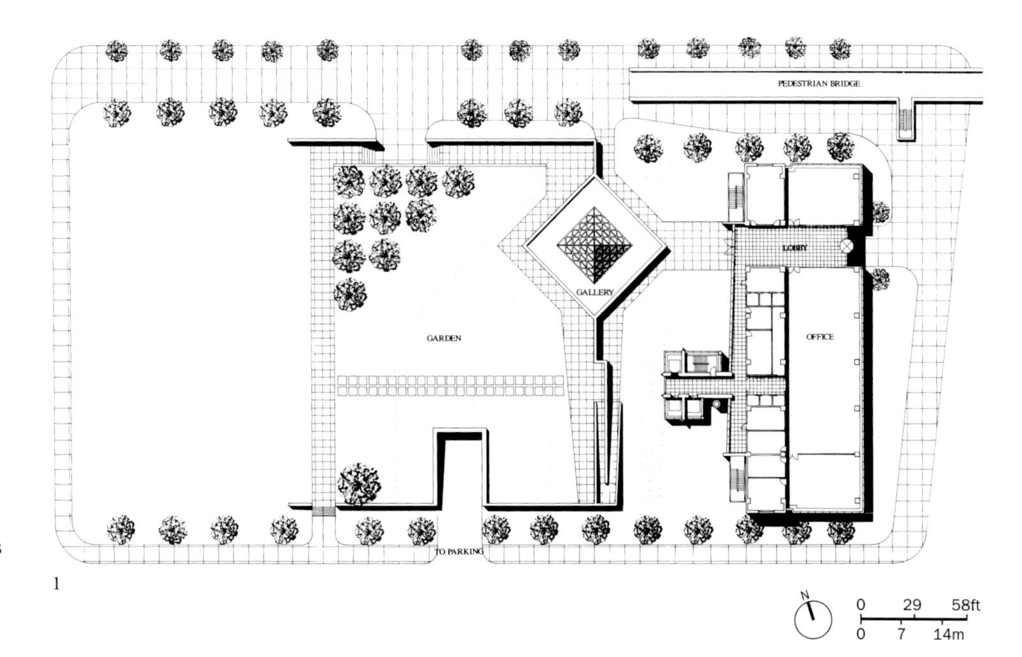

1

2

1 Site plan
2 Interior: gallery
3 Exterior at night
4 Interior: lobby of office building

3

4

Columbia Savings & Loan

Design/Completion 1988/1991
Beverly Hills, California, USA
Columbia Development Co.
83,800 square feet
Steel frame
Limestone, granite, marble, tinted glass, painted aluminum panels, copper cladding

The project involves three building sites located several blocks away from one another, south of Wilshire Boulevard. The original program called for two speculative office buildings and a headquarters building, each with a height limit of three stories and development rights twice the property area. The master plan placed the headquarters building on the center site with two additional buildings to the east and west.

The Wilshire Boulevard context dominates the design scheme, the rhythm of the street wall and building edges being legible to drivers rather than pedestrians. The project, conceived as a trilogy, allows for variation in color, scale, and composition, but unifies the buildings through the consistent use of architectural elements. The facades are composed of a set of planes on expressed structural frames, with a limestone wall forming the primary plane.

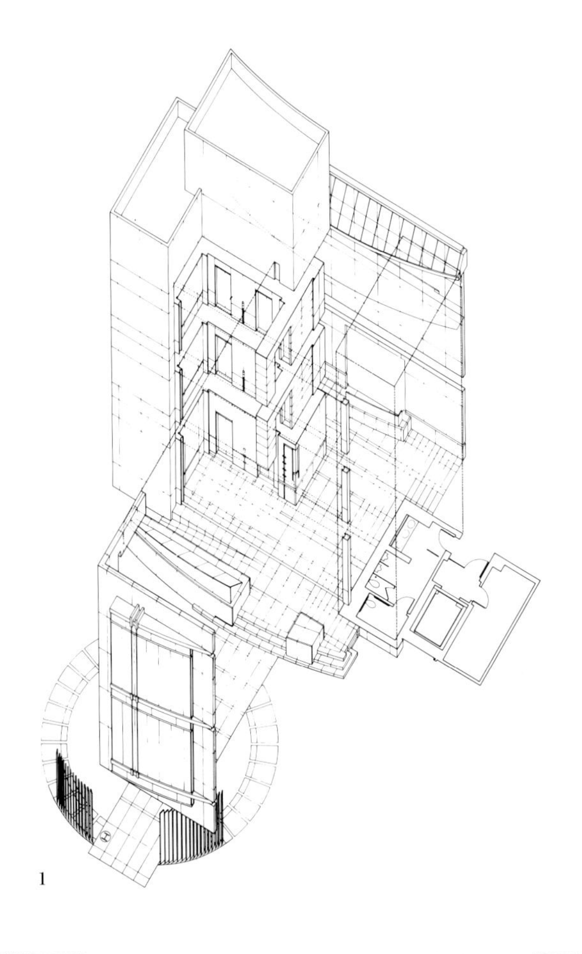

1

2

1 Axonometric of courtyard entrance to headquarters building
2 Headquarters courtyard entry
3 Entry to Wilshire and Elm

3

Asian Development Bank Headquarters

Design/Completion 1985/1988
Manila, The Philippines
Asian Development Bank
1,400,000 square feet
Cast-in-place concrete moment frame
Granite, aluminum and glass curtain wall

The project is composed of a nine-story office block and a two-story special facilities block, sited within a large landscaped area and linked by enclosed arcades. The basic plan is generated by courtyards for arrival and dining, and a central garden. Towers punctuate the mass of the office block. Floor space is planned around two large skylit atrium spaces that allow natural light into the interior offices.

In addition to offices, the building incorporates a library, a training center, a communications center, a computer facility, an auditorium, meeting and conference rooms, dining rooms and cafeterias, and a number of administrative, recreational, and service facilities.

1

1 Library atrium
2 Administrative building from central courtyard
3 Courtyard garden

2

3

Pacific Bell Administrative Complex

Design/Completion 1982/1986
San Ramon, California, USA
Pacific Bell
2,000,000 square feet (site area: 100 acres)
Steel truss
Precast concrete

Four building wings arranged in two L-shaped blocks house administrative offices, executive offices, conference facilities, dining, and utility support spaces. The wings divide the complex into four quadrants, with landscaped surface parking for 4,500 cars in three of the quadrants. The fourth quadrant is entirely landscaped, and is adjacent to an 1,800-seat dining facility. An open space at the center links two of the quadrants under a vast arched trellis between the buildings. A formal garden under the arch is typical of the landscaped areas throughout the facility.

Office interiors are based on a 50,000 square foot module that incorporates open offices and core areas, with a modest number of private offices. The large-span structure leaves the office spaces almost entirely column-free and very flexible for changing office layouts. Mullion-free, full-height glazing allows panoramic views of the landscape and the hills beyond.

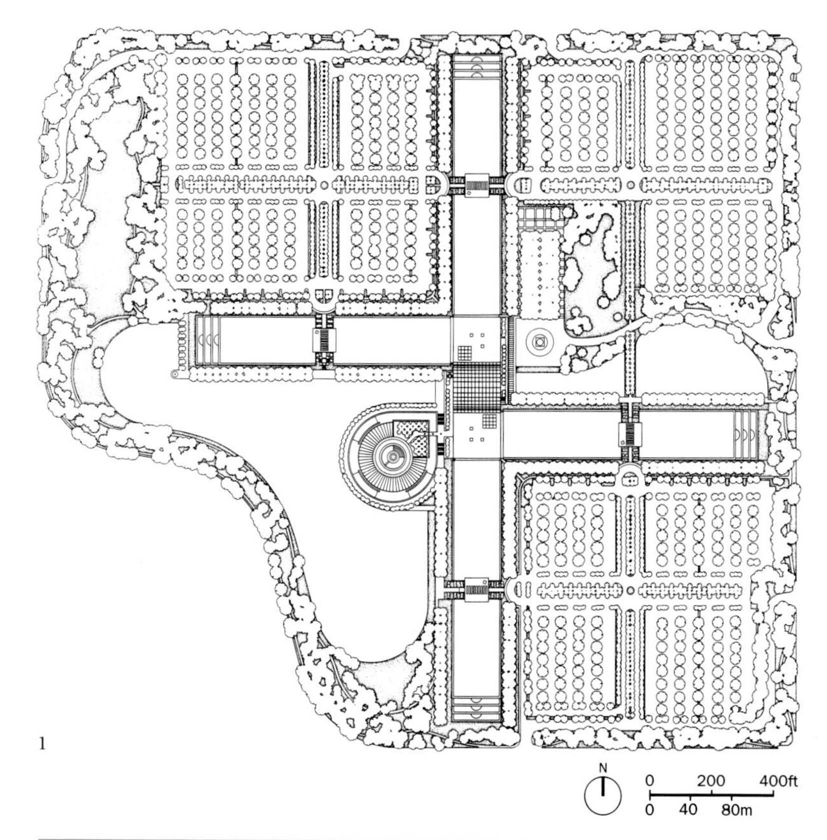

1

2

1 Site plan
2 Lakeside view from the east
3 Main lobby

3

Lucky-Goldstar

Design/Completion 1985/1987
Seoul, South Korea
Lucky-Goldstar Group
1,722,300 square feet
Steel/concrete composite structure
Luna Pearl granite

SOM provided programming services to assist in establishing design criteria for this new corporate headquarters that accommodates the merger of 23 companies. The project consists of twin 33-story towers united by a great outer court and two smaller inner courts. Because of the project's superior location, its massing and interior configurations have been carefully planned to maximize the magnificent views of the downtown area, the river, and the mountains.

A sequence of spaces provides a transition from the expansive plaza to the main building lobbies. This sequence leads from the vast open area of the plaza through the traditional gateway that marks the entry to the Lucky Building into the great outer court. From here the pedestrian is directed into the smaller inner court of either tower, these leading into the main building lobbies.

1

1 Chairman's private gallery overlooking the Hahn River
2 Secretarial waiting area
3 Chairman's private office

2

3

Latham & Watkins Law Offices

Design/Completion 1986/1989
Los Angeles, California, USA
Latham & Watkins
206,000 square feet

Latham & Watkins is located on nine floors of the First Interstate World Center, whose form is dominated by a geometry of rotating squares. The interior design is based on that geometry, with a second geometry introduced to offset its rigidity; the result is a series of curves that radiate from points within the building, softening the space and creating clear circulation patterns.

This geometric concept is reflected in the furnishings, custom-designed rugs, and decorative details. The creative use of inexpensive materials results in solutions of visual interest that allow for the graceful application of more luxurious finishes in limited areas. Attorneys' office doors are of standard aluminum frames painted in metallic finishes. Terrazzo flooring in the main reception area is detailed with aluminum divider strips in a design derived from the original geometry. Etched glass used throughout allows light into the interior spaces while providing privacy.

1

1–2 View of reception area
3 Office area

2

3

4 View of elevator lobby
5–6 View of law library

4

5

6

Sun Bank Center

Design/Completion 1985/1987
Orlando, Florida, USA
Lincoln Property Company/Sun Bank
650,000 square feet (site area: 5.9 acres)
Steel frame
Granite, marble, glass, metal details

A multi-use complex in the heart of Orlando, the project is the most substantial single development in the history of the city. The 30-story tower is connected to an existing 10-story office building via an eight-story glass-enclosed atrium and 45-foot-long bridges on selected floors. Customers pass through a two-story ceremonial entrance into a main banking hall. This is a 110-foot-high semicircular space carved from the base of the building and realized in three types of stone and a wall of lacquered teal within a grid of bronze insets.

The interior design is particularly noteworthy for its use, in a commercial space, of furnishings more readily associated with residential settings. While floors four through eight are typical administrative spaces with Knoll workstations, the upper executive floors are characterized by anigre walls, etched glass panels, custom-designed office furniture with lacquered bases and blue pearl granite tops, and a collection of Biedermeier furniture.

1

2

1 Executive office suite
2 Executive waiting area
3 Executive reception

3

4

5

4 Cafeteria
5 Servery
6 General office area
7 General office
8 Small conference room
9 Large conference room

6

7

8

9

Credit Lyonnais Offices

Design/Completion 1990/1991
New York, New York, USA
Credit Lyonnais
175,000 square feet

The new Credit Lyonnais offices establish a highly visible New York presence for a prominent European financial institution. The fixed core and adjacent corridors generate the configuration of each floor. Full-height demountable partitions create distinctly separate work units with custom office furnishings accommodating computer equipment. Star-shaped trading desks replicate those of the Paris headquarters. Throughout, a hierarchy of materials progresses from fabrics to leathers, and from laminates to wood veneers, reinforcing the functional progression from operations and data processing to banking and trading, and finally to the executive floors.

The neutrality of beige carpet, white textured walls, and light wood ceilings identifies the corridors encircling the central core. The brilliant blue of the Credit Lyonnais logo dominates fabric colors on the operational floors, while on banking floors and at areas of greater client contact, warmly colored leathers and cherry wood veneers set the tone. The project also included the renovation of the lobby and exterior.

1

2

1 Trading room
2 General office area: operations
3 General office area: banking
4 Conference room

3

4

Lehrer McGovern Bovis Inc. Offices

Design/Completion 1993
New York, New York, USA
Lehrer McGovern Bovis Inc.
46,000 square feet

The design scheme provides office space that doubles as a marketing tool and as an executive office suite for a major construction management firm.

Custom-designed workstations for teams of four can expand or contract incrementally. Private glass-walled offices are centrally grouped and integrated with the workstations, minimizing separation of senior staff. A multi-purpose room for seminars and training, as well as regulation half-court basketball, further encourages the firm's team approach.

A curved wall incorporating inset television monitors, duratrans, and an LED news jet links the work area and elevator lobby. Glass-fronted conference rooms situated in the team areas allow visitors to observe the work process.

Materials include open metal decking for the ceiling, concrete-colored carpet, and wall textures that simulate fireproofing, corrugated cardboard, and sandpaper. Wood plinths, metal angles, and I-beams make up the vocabulary of the custom-designed furniture.

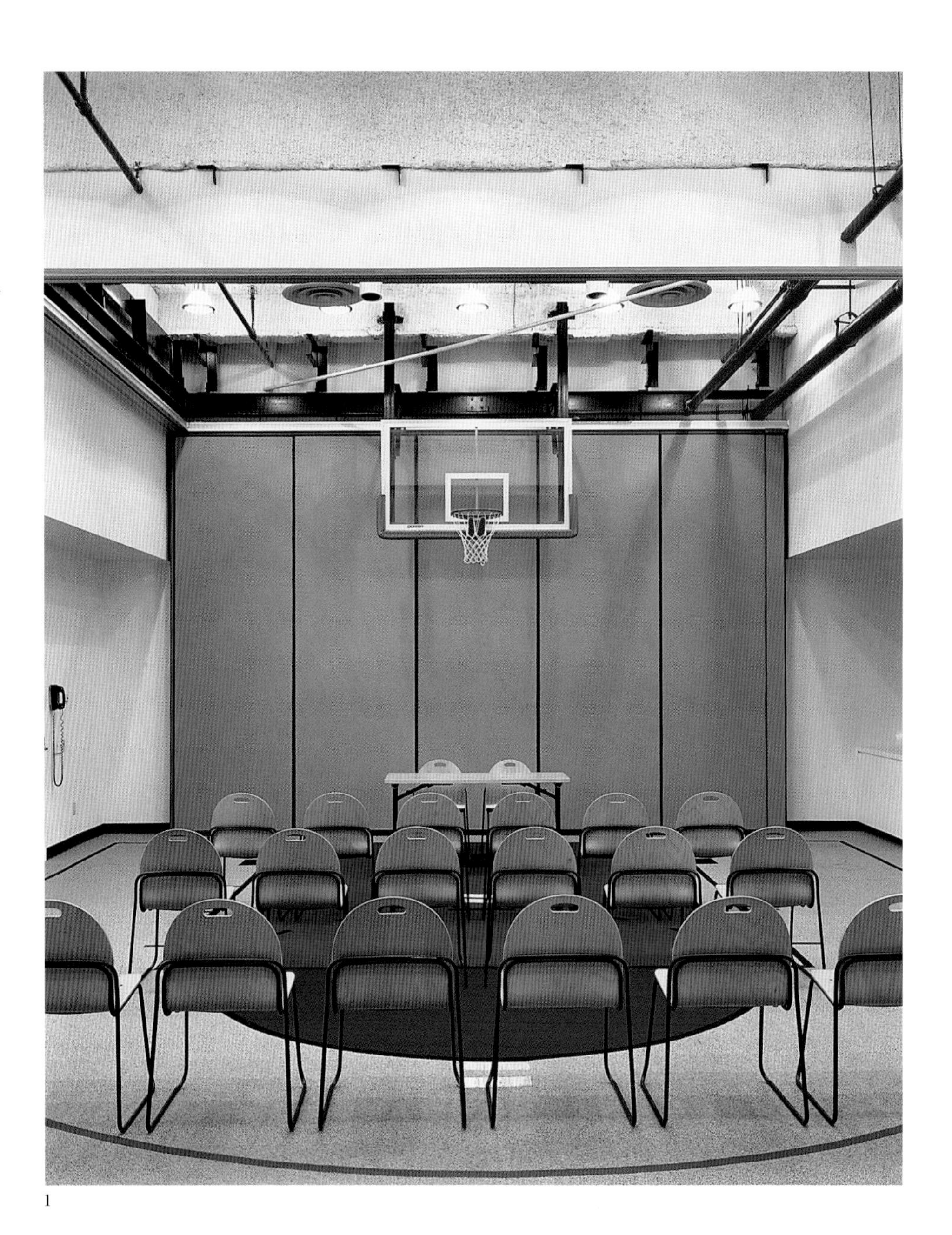

1

1 Multi-purpose room
2 Executive boardroom
3 Project team room
4 General office area
5 Typical office

2

3

4

5

Merrill Lynch Consolidation Project

Design/Completion 1986/1988
New York, New York, USA
Merrill Lynch & Company Inc.
4,250,000 square feet

A project requiring a full range of services from programming and planning to interior architecture, the consolidation relocated approximately 14,000 employees. Major programmatic elements include executive offices and reception areas; operations, administrative, and customer contact facilities; trading floors and investment banking facilities; computer business support; an auditorium; health/fitness centers; and multiple food service facilities.

Architectural elements unify the project, while materials and finish selections establish a hierarchy of spaces. Executive offices are characterized by the contemporary interpretation of traditional details, the use of antiques, and an extensive art collection. Manufactured systems furniture throughout the offices is enhanced by a number of customized pieces. The three 500-person trading rooms incorporates customized trading desks and finely tuned acoustic and mechanical systems.

1

1 Trading desk
2 Trading room

2

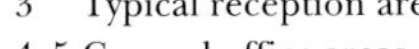
3 Typical reception area
4–5 General office areas

3

4

5

6 Rotunda
7 Executive corridor
8 President's office

6

7

8

Selected and Current Works

Design for Pubilc Use

Canary Wharf

Design/Completion 1985/1991
London, England
Olympia and York Canary Wharf Ltd
12,405,000 square feet (site area: 71 acres)

Designed to accommodate the City's expanding financial activities, the development revitalizes a vacated Port of London site, providing office and trading floor space, as well as public recreational and retail facilities. SOM provided development guidelines, as well as designing infrastructure elements including roads, parks, gardens, and other open spaces.

Canary Wharf consists of 26 separate building sites, the majority of which are located over the water and along the perimeter of the wharf. The proposed three landmark towers will be visible from a considerable distance. A number of mid-rise buildings are grouped around the public spaces to frame the sequence of landscaped areas. To maximize development, the buildings were constructed over the water at points to the north and south of the wharf. As a result, the river basins no longer form a linear passage, but are a series of water courts ringed by pedestrian areas.

1

1 Axial view from the west
2 Master plan of Canary Wharf
3 View within the retail arcades off Cabot Square

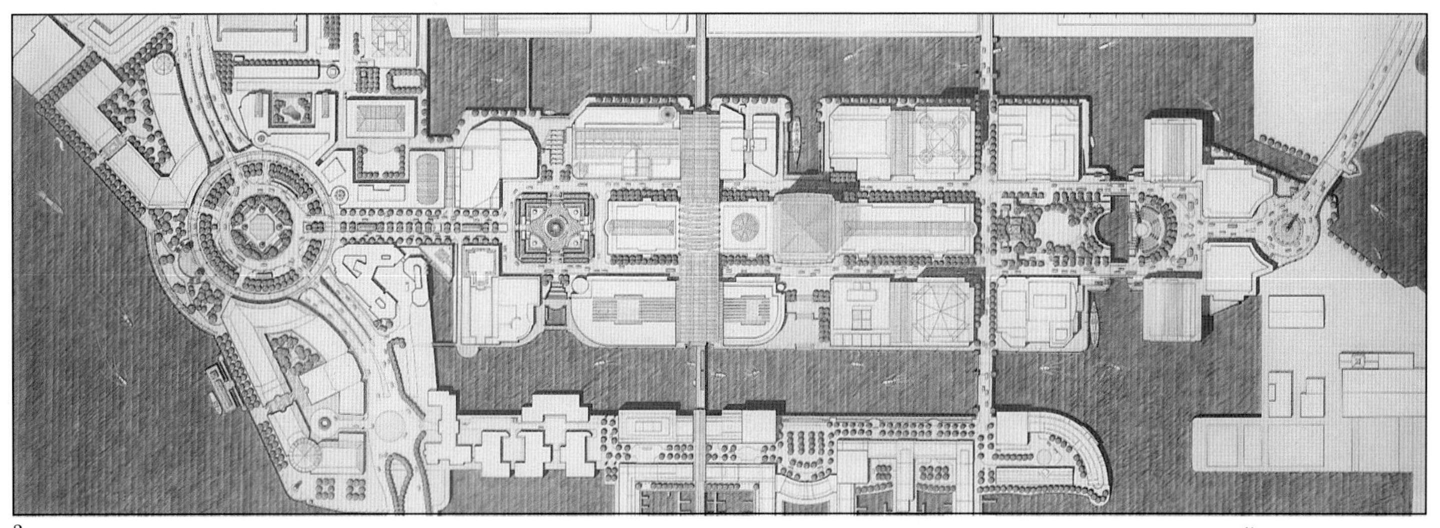

2

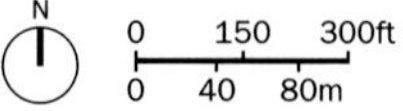

3

4 View towards the central space, Cabot Square
5 View along the wharf promenade
6 View from the tower over Cabot Square

4

5

6

Riverside South Master Plan

Design/Completion 1992
New York, New York, USA
Riverside South Planning Corporation
6.1 million square feet (site area: 55 acres)

Situated along the Hudson River between 59th and 72nd Streets, Riverside South is a tract of undeveloped land. Formerly the site of an extensive rail yard, the area is still occupied by an active Amtrak line, a portion of the elevated West Side Highway, and inaccessible, dilapidated piers. The plan brings the city back to the river by extending the street grid westward and introducing a public park along the waterfront.

The development program consists of 5,700 residential units, neighborhood retail, offices, cultural and community facilities, film studios, and a 23-acre riverfront park. A crucial element of the program is the relocation of the West Side Highway inland and at-grade. The master plan prescribes land use, building envelopes, urban design guidelines, an open space plan, and a phasing plan. The neighborhood vocabulary of townhouses, streetwall apartment buildings, and narrow towers characterizes the planned buildings.

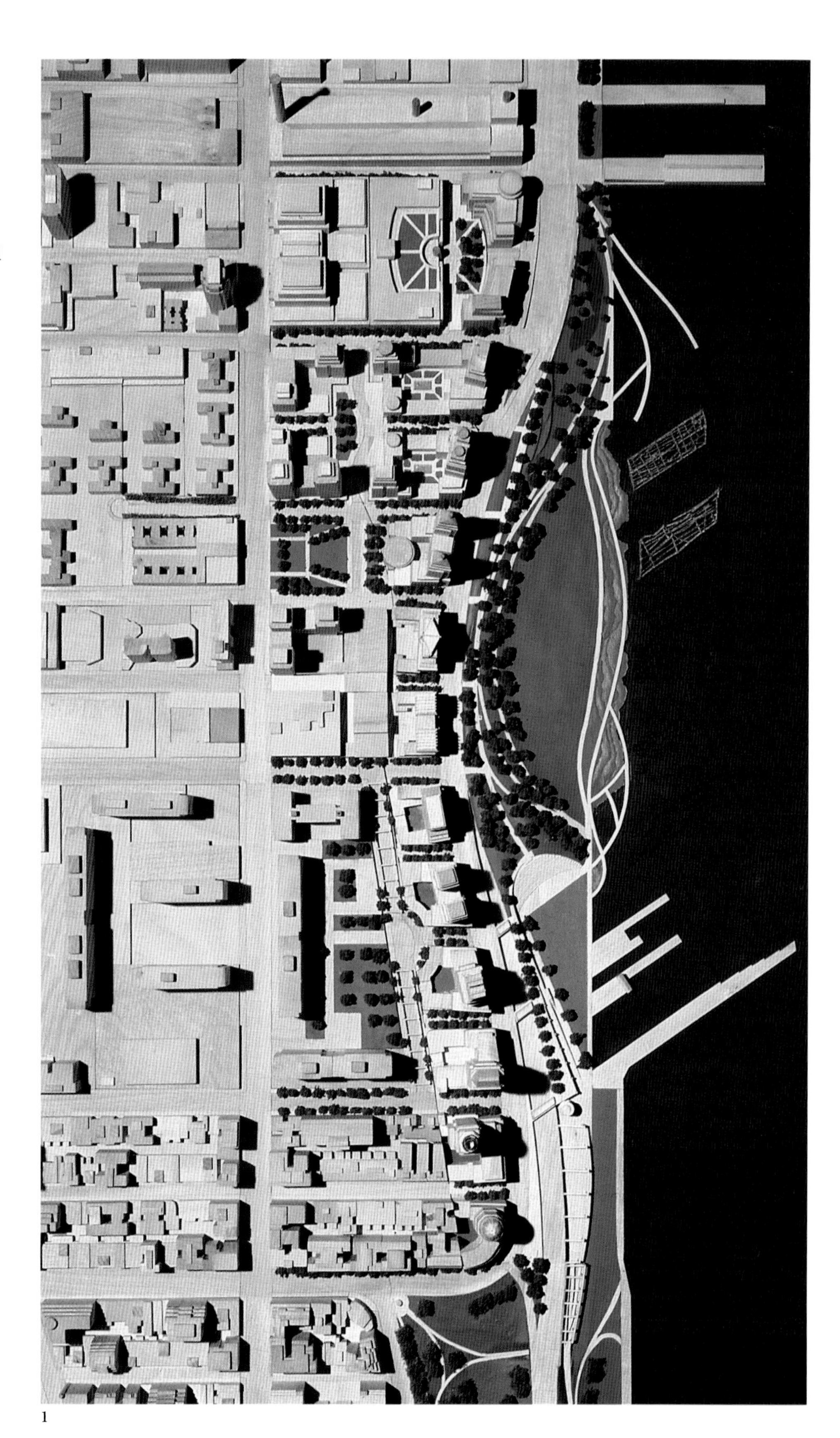

1

1 Model: site plan view
2 Rendering of project from the north

2

Mission Bay Master Plan

Design/Completion 1987/ongoing
San Francisco, California, USA
Catellus Development Corporation
Site area: 315 acres

A comprehensive master plan for the mixed-use development of a 315-acre site south of downtown San Francisco, the project is located along the waterfront, creating a new district on land formerly owned by the Southern Pacific Railroad. The centerpiece of the plan is an integrated community which will become an economic and housing resource for the city. Approximately 8,000 residential units are planned to house people of varying incomes, including singles, families, and individuals with special needs.

Mission Bay is a predominantly pedestrian environment, with parks and open spaces marking major commercial centers and providing view corridors and pedestrian connections to the waterfront and adjoining districts. When completed, Mission Bay will be one of the largest privately sponsored urban developments in California.

1

2

3

1 Long Bridge Plaza: strolling towards the bay
2 Afternoon on King Street, Mission Bay's commercial boulevard
3 An outdoor cafe in Crescent Park
4 Aerial view of Mission Bay
5 Site plan

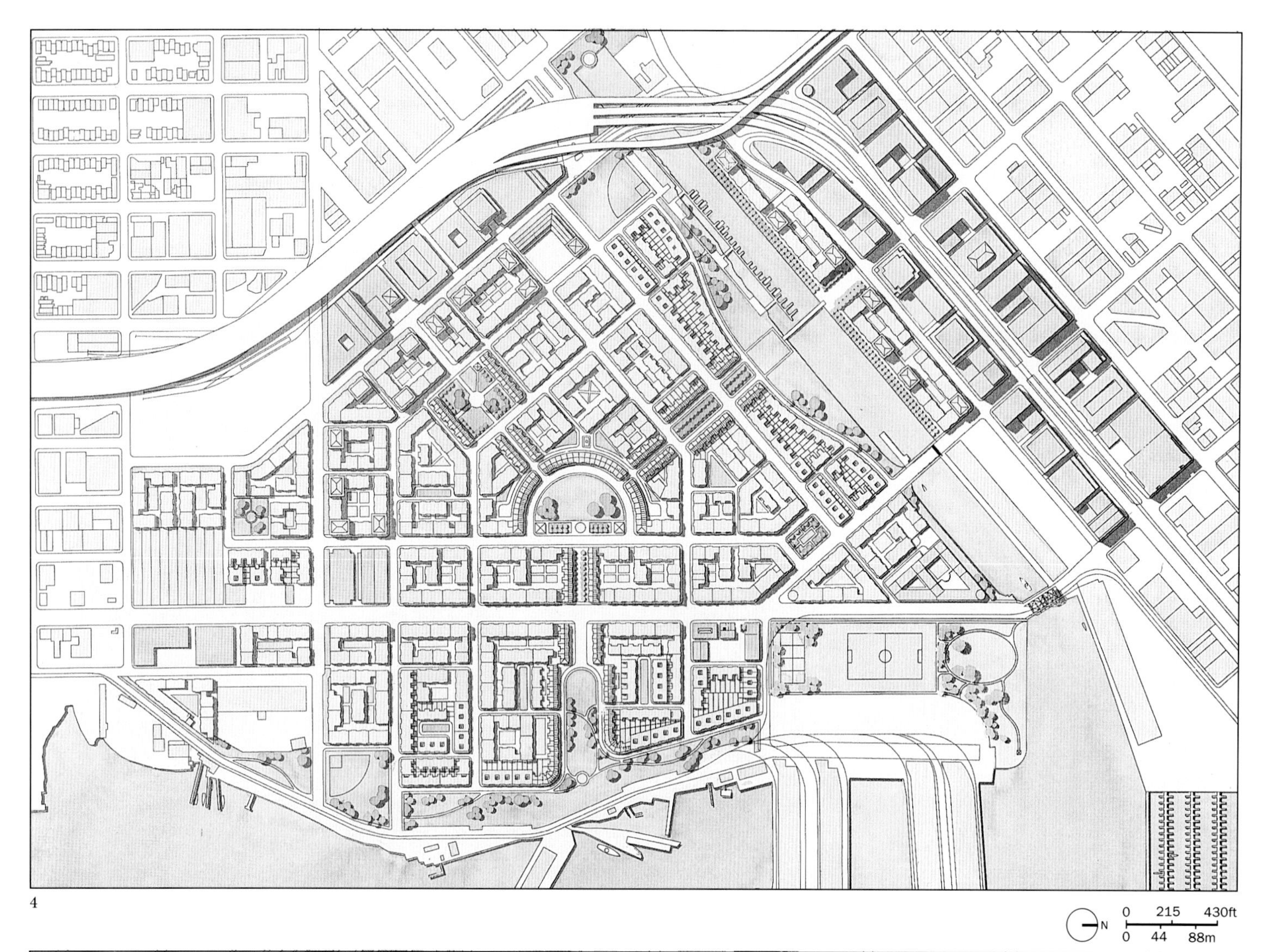

4

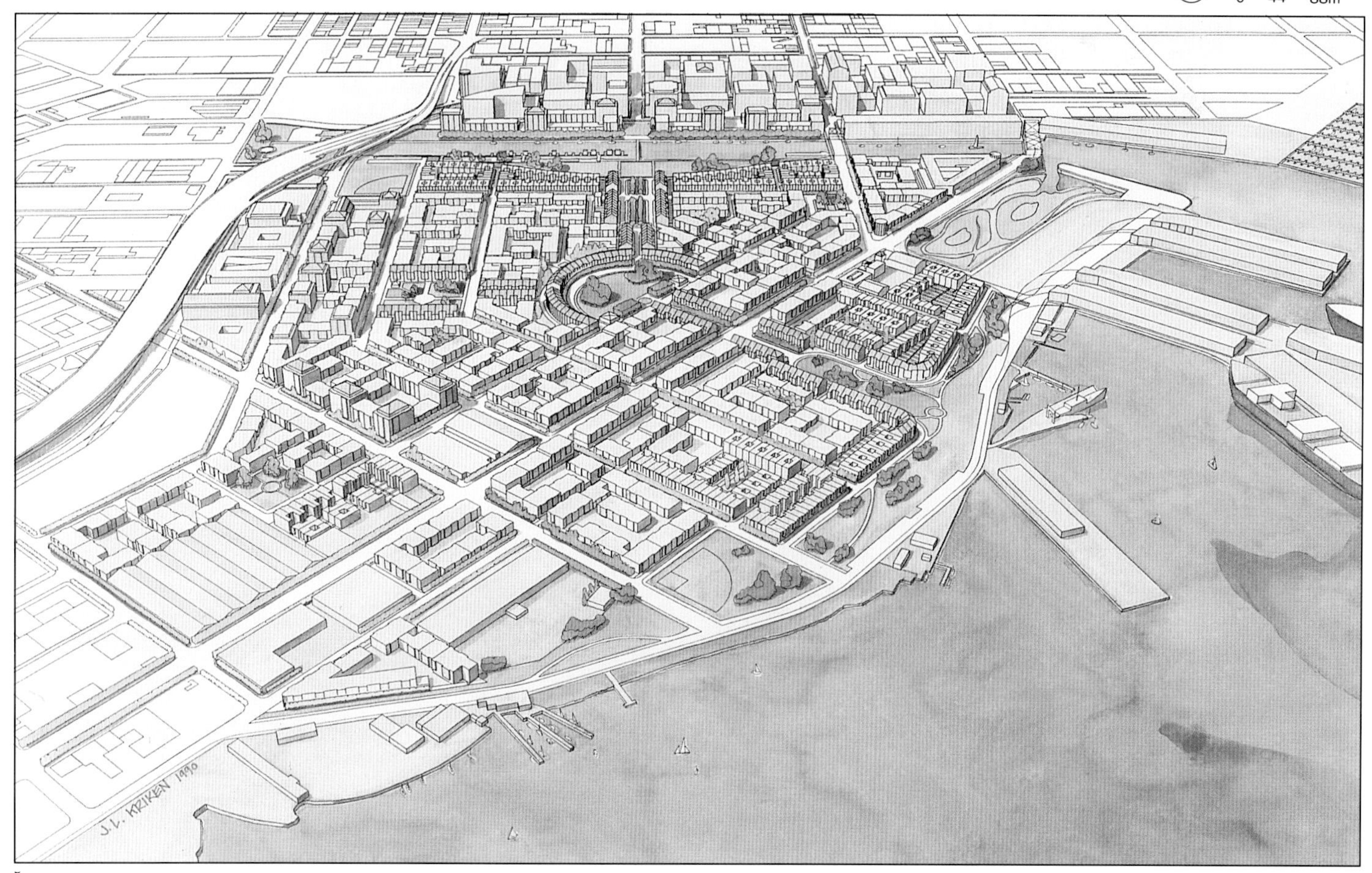

5

Saigon South

Design/Completion 1993/ongoing
Ho Chi Minh City, Vietnam
Central Trading and Development Corporation

Saigon South extends for 11 miles along a new roadway that connects a new deepwater port and export processing zone to the regional road system and international airport. The centerpiece of the project is the 988-acre mixed-use New City Center, planned to accommodate over 100,000 people by the year 2000. Future projects will include research and development facilities for high-tech industry, several universities, a regional medical center, a sports stadium, and community parks and open space.

The planning concept of a "city of islands" preserves the district's many existing canals and landscape features as a framework for development. At the same time it allows for incremental growth that provides a sense of completion at every phase.

1

2

3

1 Entering Saigon South via the new roadway
2 The Crescent Promenade creates a sense of identity for the Crescent District
3 Waterfront parks frame the arrival by water-taxi to the city's Central District
4 Saigon South is a city of islands
5 The New City Center

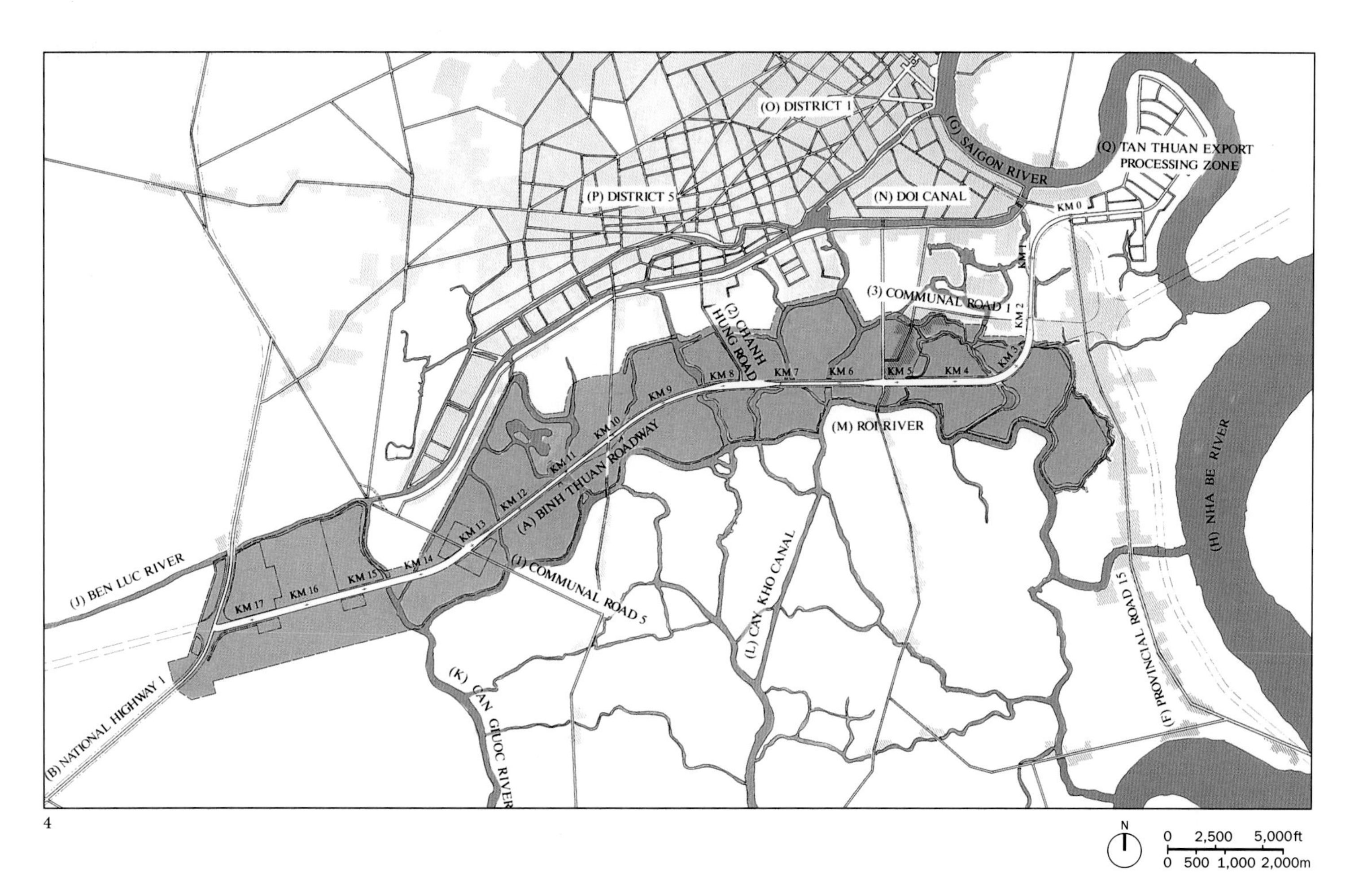
(O) DISTRICT 1
(G) SAIGON RIVER
(Q) TAN THUAN EXPORT PROCESSING ZONE
(P) DISTRICT 5
(N) DOI CANAL
KM 0
KM 1
KM 2
(3) COMMUNAL ROAD 1
(2) CHANH HUNG ROAD
KM 3
KM 4
KM 5
KM 6
KM 7
KM 8
KM 9
KM 10
KM 11
KM 12
KM 13
KM 14
KM 15
KM 16
KM 17
(M) ROI RIVER
(A) BINH THUAN ROADWAY
(J) BEN LUC RIVER
(1) COMMUNAL ROAD 5
(L) CAY KHO CANAL
(K) CAN GIUOC RIVER
(B) NATIONAL HIGHWAY 1
(F) PROVINCIAL ROAD 15
(H) NHA BE RIVER
N
0 2,500 5,000 ft
0 500 1,000 2,000 m

4

5

Potsdam

Design/Completion 1992/2000
Potsdam, Germany
LEG
Site area: 4,300,000 square feet (99 acres)

This project involves a feasibility study and master plan for the redevelopment of rail yards and derelict industrial land adjacent to the historic city center of Potsdam.

The master plan proposals include commercial development bridging the railway tracks which currently divide the site. Phased redevelopment involves initial site reclamation and construction beginning with a new rail terminal, with commercial development on the western portion of the site. Subsequent phases of construction will include a hotel, residential accommodation and additional office buildings in a mixed arrangement across the site. New roads and utility infrastructure will be provided and a riverside marina will act as the focus for the new residential area.

1

2

3

1 View from riverside marina to Nickolei Kirche
2 View to listed water tower
3 View of typical landscaped plaza
4 Model: aerial view from the south
5 Site plan of proposed development

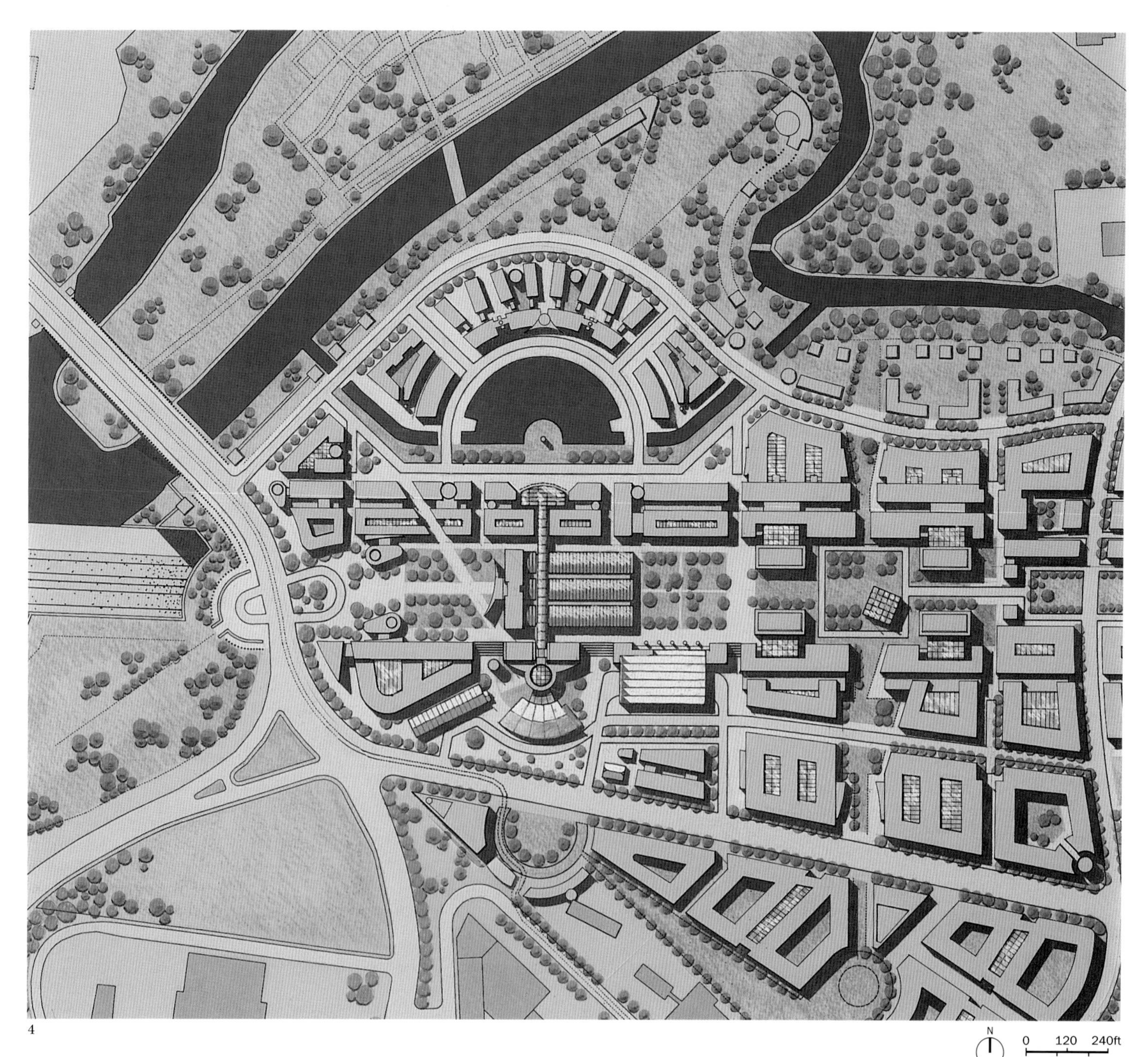

4

5

Doncaster Leisure Park

Design/Completion 1992/2002
Doncaster, Yorkshire, England
Doncaster Metropolitan Borough Council
Site area: 300 acres

This leisure and business park project is located 1.5 miles east of Doncaster town center on 300 acres of derelict land, part of which is a disused airfield. Adjacent to the site are residential communities, new industrial development, a nature reserve, a new recreation center, and Doncaster Racecourse.

The master plan design centers on a 50-acre lake which serves as a visual focus for the entire development, including business park facilities, leisure facilities, and an expanded residential district accommodating 300 new housing units in a variety of building types with ancillary retail and community services.

A critical component of the design is the extensively landscaped parkland, which reuses poor-quality material found within the site, as well as excess material from the lake excavation. This creates a hilly area that provides shelter from prevailing winds, as well as visual interest for residents and visitors alike.

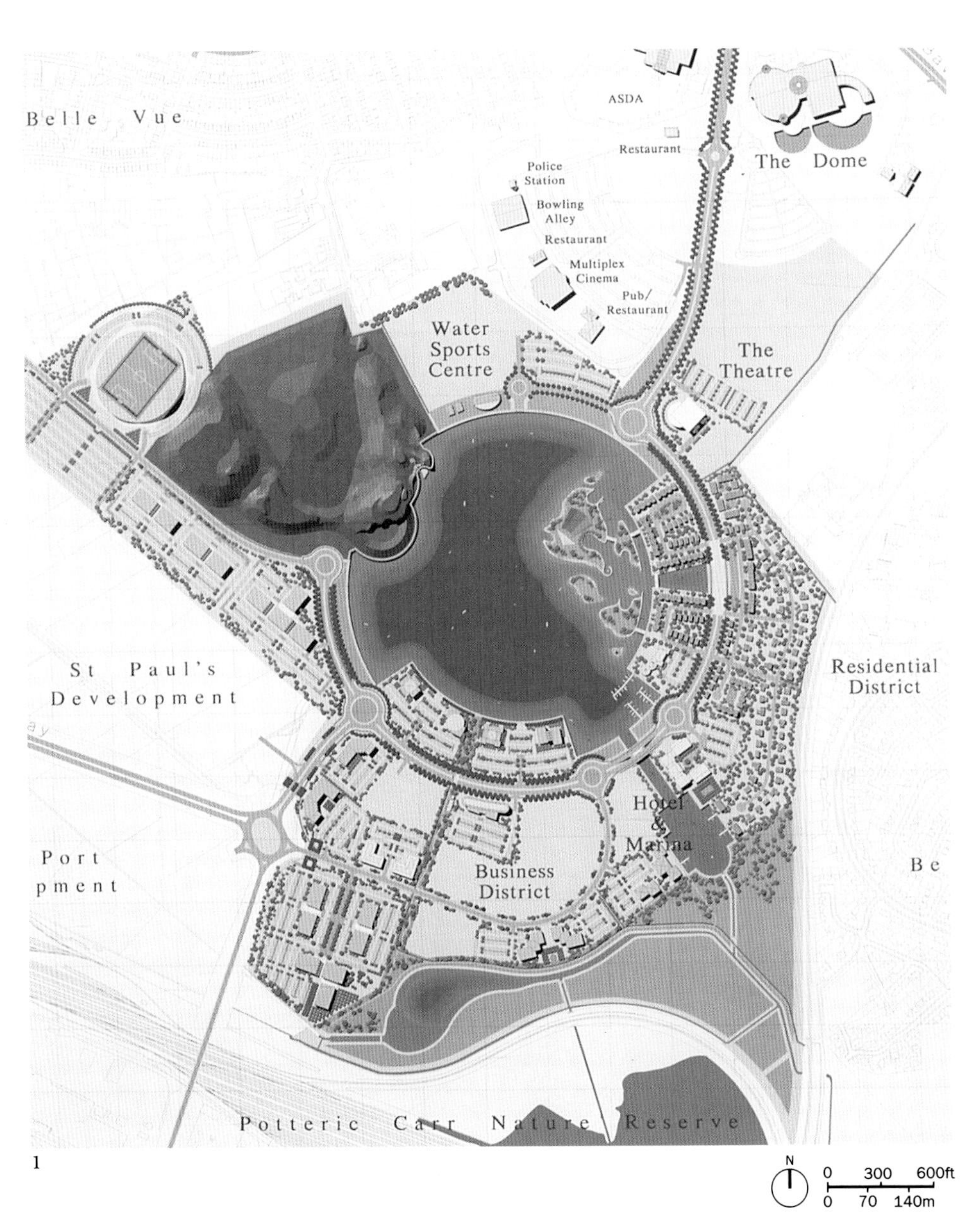

1

2

1 Illustrative master plan
2 Computer rendering: lake and island features
3 Computer rendering: aerial view from the south
4 Computer rendering: business district

3

4

Sydney Harbour Casino

Design/Completion 1993
Sydney, Australia
Circus-Circus
Site area: 11 acres
Reinforced concrete and steel frame
Precast stone, glass panels, steel-framed vision glass

The waterfront complex is a multi-use development including commercial, retail, and dining facilities; gaming halls; a 600-room hotel with meeting rooms, a health club and ballroom; a 7,050-car garage below a new park; a series of indoor and outdoor recreational facilities; and a marina.

Surrounded by a residential neighborhood and a declining industrial community overlooking Sydney Harbour and the central business district, the project introduces a neighborhood park with casino and parking facilities below. The program distributes components across two former industrial piers and adjacent vacant waterfront land. The strategy of merging land and water activities continues in the multi-terraced waterside pedestrian promenades, shops, restaurants, docks and boating slips, and sports and entertainment facilities.

The Sydney Harbour Casino is designed to serve both the city of Sydney and the world as a 24-hour family entertainment experience during the 2000 Olympic Games.

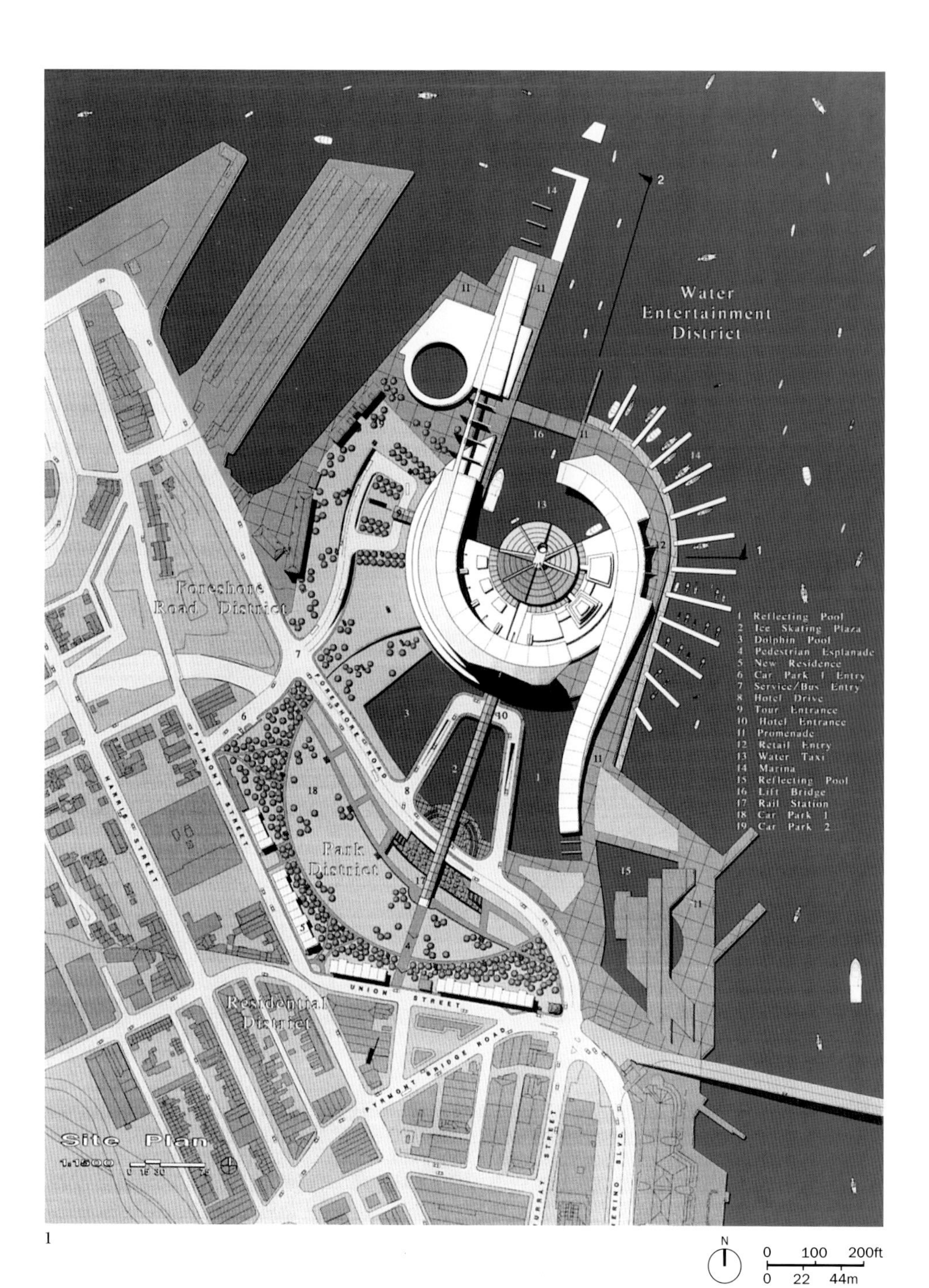

1

1 Site plan
2 Harbor view looking south-east
3 Aerial view to the south

2

3

Broadgate Leisure Club

Design/Completion 1988/1990
London, England
Rosehaugh Stanhope Developments PLC
21,520 square feet

As part of the Broadgate development at Liverpool Street Station, the Broadgate Leisure Club provides a mix of leisure, health and sports facilities for members of the local business community. Facilities include a bar, restaurant, and swimming pool, as well as squash courts, exercise rooms and a comprehensive gymnasium.

The design focuses on the swimming pool which is located in a large, double-height hall adorned by a 100-foot mosaic wall designed by Howard Hodgekin. Although the club has no direct exposure to the outside, the design creates the illusion of a spacious outdoor environment.

The club is on two levels, with the reception lobby, lounge, and bar areas above, and sports areas and changing facilities below. The staircase is an extension of the public areas with landings and intermediate spaces that accommodate tables and chairs, allowing views of the squash courts below.

1 Swimming pool, training rooms, and restaurant
2 Reception area and restaurant
3 Swimming pool with artwork wall

2

3

Arlington International Racecourse

Design/Completion 1987/1989
Arlington Heights, Illinois, USA
Duchossois Interests
Site area: 350 acres
Steel frame with cantilevered steel truss roof
Split-face limestone, painted brick, white painted metal panels, glass

Built to replace the original grandstand after its destruction by fire in 1985, the project redefines the concept of a traditional racing center by increasing visitor accessibility to the full range of activities that support horse racing. The new Paddock Garden is the focal point, placed between the grandstand and a group of smaller buildings for stalls, saddling and grooming, jockeys, and related functions. Pre-race ceremonies of saddling, mounting, and walking the horses can be viewed from beneath garden trees, as well as from the grandstand balconies and arcades overlooking the paddock area. These features are unique to Arlington, as most grandstands face the racetrack only.

The low-rise section, with its cantilevered steel truss roof, provides unobstructed views from all vantage points.

1

2

1 Clubhouse entrance
2 Grandstand viewed from the paddock
3 View of the track from the restaurants on level 4

3

Hubert H. Humphrey Metrodome

Design/Completion 1977/1982
Minneapolis, Minnesota, USA
Hubert H. Humphrey Metrodome
Associate architect: Setter, Leach & Linstrom Inc.
63,000 square feet football/soccer; 56,360 square feet baseball; 450,000 gross square foot enclosure
Cast-in-place and precast concrete
Teflon-coated translucent fiberglass roof

The project provides facilities for several professional sports in an enclosed structure, seating up to 62,000 for Vikings football and Kicks soccer games, and up to 55,000 for Twins baseball games. Additional seating is provided in 115 private suites and on-field seating is provided for entertainment events. Team rooms and service areas are located on two partial levels below grade.

The 340-ton roof comprises two layers of teflon-coated translucent fiberglass fabric restrained and shaped by a network of 18 steel cables anchored to a perimeter compression ring. While the outer layer of this fabric is completely airtight, the inner layer acts as an acoustic baffle, creating a thermal barrier and a plenum for the movement of warm air under the roof. Because air pressure must be maintained for roof support, all entrances are revolving doors.

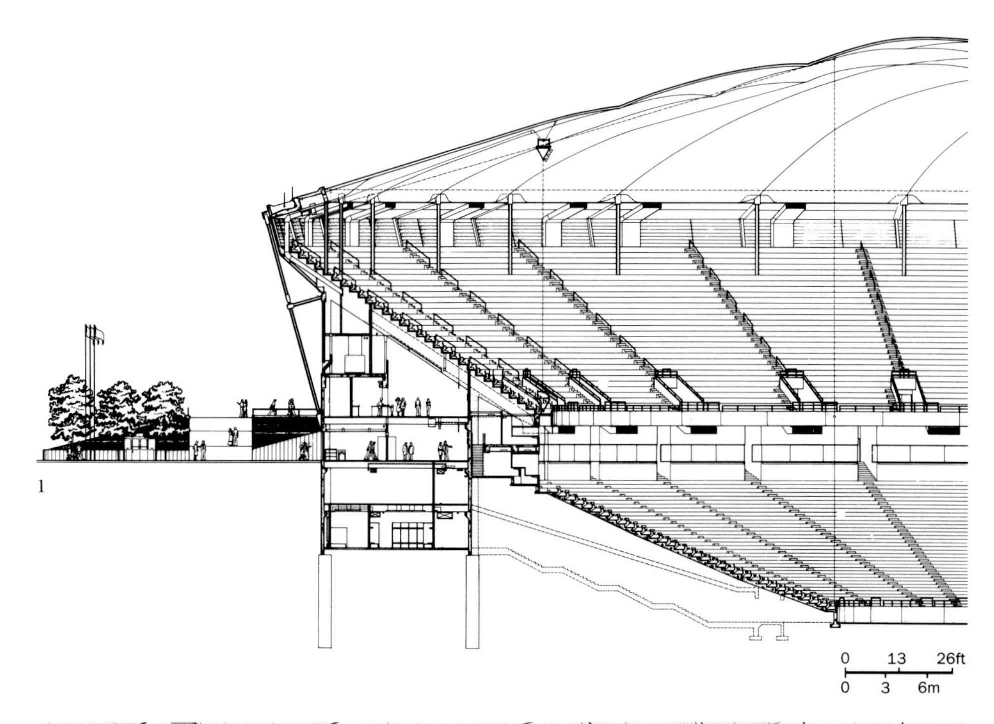

1

2

1 Section
2 Interior view during baseball game
3 View from the south

3

Utopia Pavilion Lisbon Expo '98

Competition 1994
Lisbon, Portugal
Parque Expo '98
323,000 square feet (17,500-spectator arena)
Laminated wood
Aluminum roof

The competition-winning scheme is a multi-purpose arena including a six-lane 200 meter Olympic track, that will be capable of accommodating the full range of sports activities as well as a variety of other events ranging from rock concerts to conferences and exhibitions. It is located in the new northern development district of Lisbon within the World Fair Expo 1998.

The arena relates metaphorically to the commemoration of the 500th anniversary of the Portuguese discovery of the route to India from Europe. The interior of the building recreates the image of the hull of a ship using 390-foot spans of laminated wood portals which are covered by a timber ceiling to generate a warm interior atmosphere.

The massing of the building has purposely been kept to a very low profile so that it does not compete with the surrounding buildings. The resulting aerodynamic form relates to the history of the site, which used to be a landing dock for seaplanes.

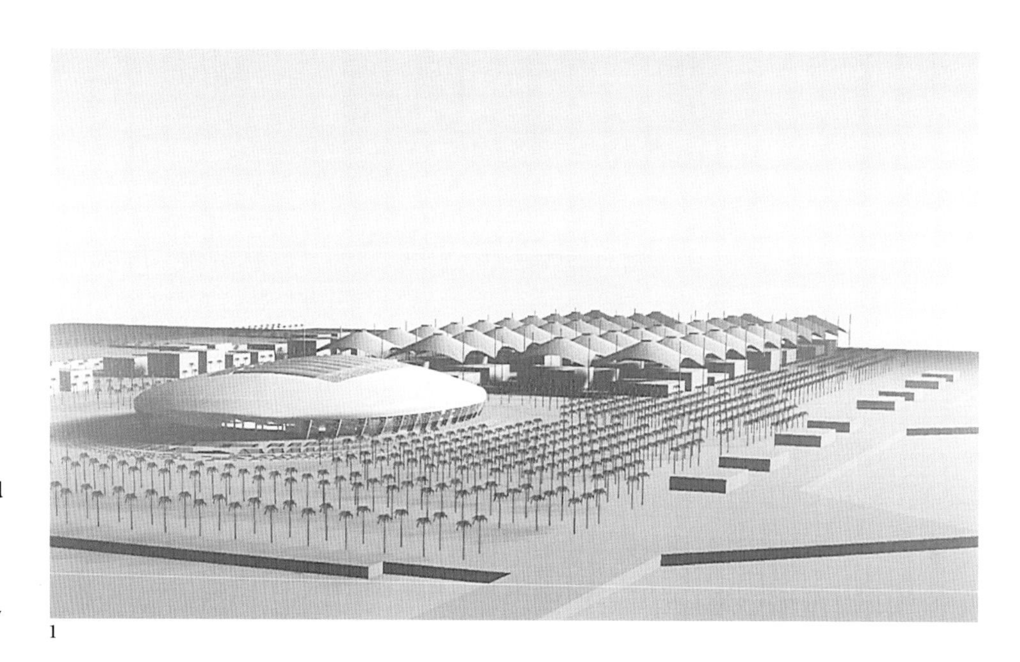

1

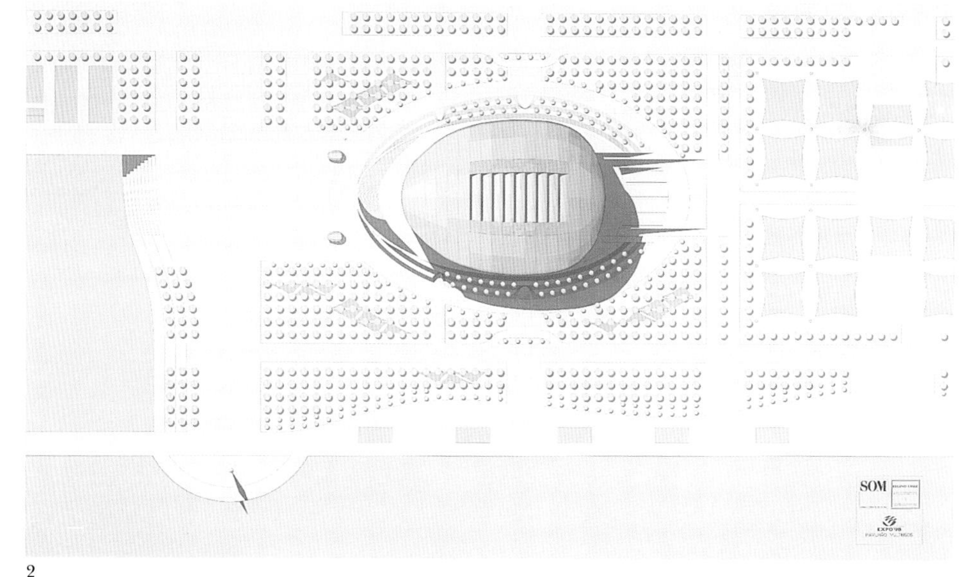

2

1 Perspective view of arena
2 Roof plan
3 Entry level plan
4 Cross section
5 Longitudinal section model

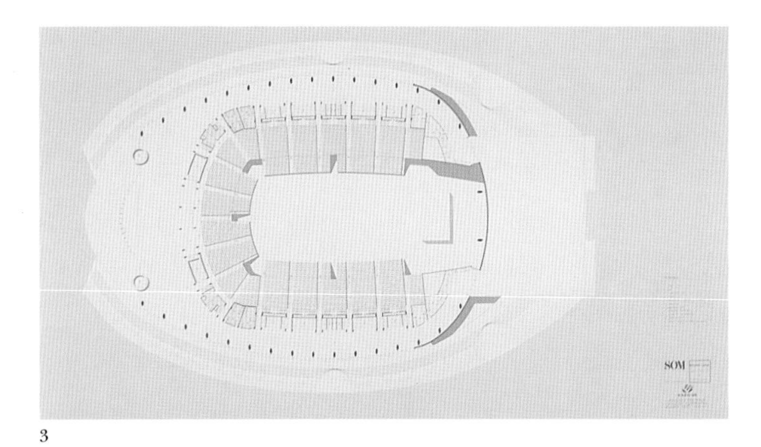

3

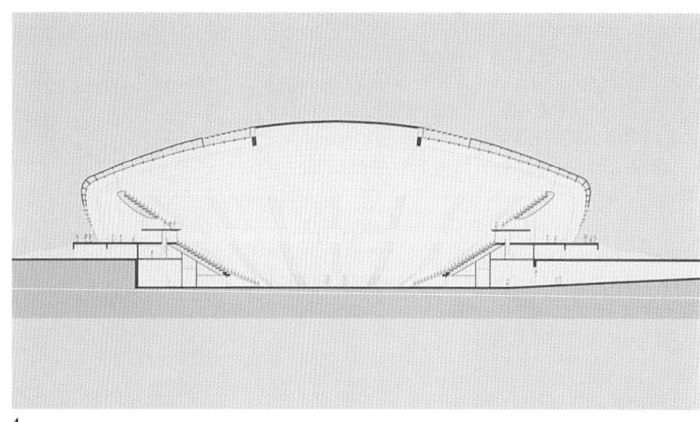
4

5

Palio Restaurant

Design/Completion 1986
New York, New York, USA
Equitable Investment Corporation
10,000 square feet

The main entrance to the Palio is through the ground-level bar, a grand room richly paneled with European bog oak. The bar itself is a slab of dark Roncevalles marble with a honed finish set above chrome-framed, stainless steel mesh panels. The slats of the bar chairs and the staccato pattern of the black and white marble floor echo similar motifs in Sandro Chia's Neo-Expressionist mural over the dado, which depicts the medieval Sienese race after which the restaurant is named.

The dining room is reached by private elevator and seats 120 with two private rooms for 40. A decorative scheme that includes Arts and Crafts-inspired wooden screens, translucent honey onyx panels, and SOM's Andover chairs creates a sedate and elegant setting enhanced by appointments designed by Massimo Vignelli. Vividly colored heraldic symbols representing the 17 *contrade*, or wards, of the City of Siena hang on the walls.

1

1 Dining room
2 Main dining room
3 Bar lounge

2

3

Chicago Place

Design/Completion 1988/1991
Chicago, Illinois, USA
Brookfield Development Inc.
365,000 square feet
Steel frame
Granite, precast concrete panels, limestone, poured-in-place concrete

This mixed-use complex consists of an eight-story retail mall and a 43-story, 272-unit apartment tower. Its design follows the building height guidelines of the 1918 architectural master plan for the area, with the tower set back 120 feet from the Michigan Avenue property line.

The retail component is organized around a 40-foot high, two-and-a-half-story lobby at the Michigan Avenue entrance and an eight-story atrium. A palm garden on the eighth floor features a barrel-vaulted glass roof supported by structural steel arches.

Internal functions are expressed on the exterior, relating the building to its downtown Chicago context. The consistent language of color and ornament creates an exciting shopping and dining environment.

1

1 Main entrance in Michigan Avenue
2 Palm garden on level 8
3 Atrium
4 View towards the main entrance
5 Atrium

2

3

4

5

Solana Marriott Hotel

Design/Completion 1988/1990
Westlake, Texas, USA
Maguire Thomas Partners and IBM Corporation
185,000 square feet
Concrete frame
Limestone, plaster, glass

The 193-room hotel incorporates two boardrooms, a ballroom, a restaurant, a bar, and a living room with fireplace and grand piano, with interiors that are designed to respond to the strong forms and colors of the exterior architecture. The architectural language is established by the stone wedge and curved ceiling of the main entry, the asymmetrical cone over the bar, and the vaulted pre-function area. Finishes of varied textures and a rich color palette create an environment of distinctly separate spaces that are unified by glimpses of one from the other.

Furnishings are chosen to enhance the mood of each space, many of them custom-designed for the hotel, with references to the Arts and Crafts Movement in their simplicity, use of natural materials, and hand-crafted construction. Carpets throughout the hotel are custom designs intended to link architectural elements with more intimately scaled furnishings.

1

1 View of lounge
2 View of registration desk
3 View of registration lobby

2

3

4 View of lounge
5 View of executive conference room facilities
6 View of lobby entrance

4

5

6

Daiei Twin Dome Hotel

Design/Completion 1990
Fukuoka, Japan
Daiei Corporation
Associate architect: Nihon Sekkei
1,000,000 square feet
Structural steel frame
Metal, stone, and glass cladding

The project consists of a hotel with convention facilities and a large retail center, and is part of the Daiei Twin Dome complex—a sports and entertainment area that is a focal point of the city's waterfront.

The design extends the rough stone base of the adjoining entertainment dome to create a platform for the new development. The elliptical 35-story hotel tower plays off the curve of the dome, its height and form resembling a sail along the water. Convention facilities are located in a separate, lower building, connected to the hotel by a retail winter garden. A pavilion and courtyard overlooking the harbor provide a dramatic outdoor setting for weddings on the rooftop of the convention hall.

1

1 Evening view of hotel tower
2 View from the Eastern Channel
3 Aerial view of hotel and convention center complex

2

3

Sheraton Palace Hotel

Design/Completion 1987/1991
San Francisco, California, USA
Kyo-Ya Company Ltd
580,000 square feet
Steel frame with masonry infill; upgraded for seismic requirements with concrete shear walls
Brick masonry infill

The program called for the restoration of original historic elements and the structural upgrading of the building. Originally constructed in 1908 in the Beaux-Arts style reminiscent of the grand hotels and casinos of the French Riviera, the nine-story hotel features 550 guest rooms and suites, and remained virtually unchanged until it was closed in 1989 for restoration. Accurate restoration of both interior and exterior materials, including stone, marble, wood and plaster, are central to the character of the project. As part of the effort, the 70,000-pane Garden Court skylight was dismantled and restored off-site.

New construction replaces 1930s rooftop additions with 45,000 square feet of meeting rooms, a skylit pool, and a health club. Custom-designed interior features include carpets, light fixtures, plumbing fixtures, and furnishings. The hotel is now in complete compliance with ADA requirements.

1

2

1 Evening view of main lobby
2 Grand corridor
3 Historic Garden Court

3

McCormick Place Exposition Center Expansion

Design/Completion 1984/1986
Chicago, Illinois, USA
McCormick Place Exposition Authority (MPEA)
1,600,000 square feet
Concrete frame
Metal panel cladding system, concrete pylons, stainless steel, glass

The project accommodates two exhibition halls, meeting rooms, administrative offices, enclosed crate storage, truck docks, and support and building services. The new building is an air-rights building sited adjacent to the original SOM-designed building, over existing railroad tracks. The components are connected by a pedestrian spine.

The main exhibition hall is clad with silver-gray metal panels attached to polished stainless steel mullions, with a band of louvers for fresh air supply. A line of vision glass just below the roof emphasizes the structural nature of the cable system above. A pattern of diagonal panels above the glass strip delineates the roof's structural truss system.

The main hall is characterized by a system of 15-foot trusses suspended from cables that are in turn hung from 12 pylons spaced to provide flexibility for exhibition configurations. The pylons also serve as air supply ducts, keeping the upper hall free from ductwork.

1

1 Detail: exterior wall
2 South elevation
3 Detail: cables and pylons

2

3

The Art Institute of Chicago
Renovation of Second Floor Galleries of European Art

Design/Completion 1983/1986
Chicago, Illinois, USA
The Art Institute of Chicago
40,000 square feet

Restoring the logic and clarity of the original 1892 plan by Shepley, Rutan & Coolidge, the renovation replaces the maze-like circulation pattern imposed by earlier remodelings with a series of corridor galleries. These artificially lit spaces are available for showing works on paper that can be displayed in proximity to paintings and sculpture by the same artists in the larger, skylit galleries. New skylights and translucent laylights in the major gallery spaces restrict harmful ultraviolet light while allowing for the natural light that enhances viewing.

The corridor galleries originally formed a rectangle in the center of the second floor, providing long, dramatic vistas and a clear circulation pattern. Walls blocking portions of the corridor were removed, and doorways repositioned according to the original plan. New glass walls were added, allowing a view of the grand entrance staircase.

1

2

1 View of light court
2–6 Galleries

3

4

5

6

Holy Angels Church

Design/Completion 1989/1991
Chicago, Illinois, USA
Holy Angels Parish
18,400 square feet
Reinforced concrete and steel frame
Masonry, glass, steel, solar panels

The new church replaces one destroyed by fire in 1986, and has been reoriented to face east, with access to the remaining rectory building. The chapel tower anchors the basilica plan at the north-east, while the glazed sanctuary and baptistry mark the west. Steel roof trusses characterize the nave which ends in a burst of light above the altar.

The pitched roof carries a glass and steel envelope the length of the south side, capturing solar energy that is drawn through a hollow concrete floor providing direct and indirect heating and cooling. Conceived as an energy-efficient thermos using component building systems, the design minimizes construction and operating costs. The main entrance is the major opening in the north wall, facing the street, while the south side is punctuated with glazed doorways.

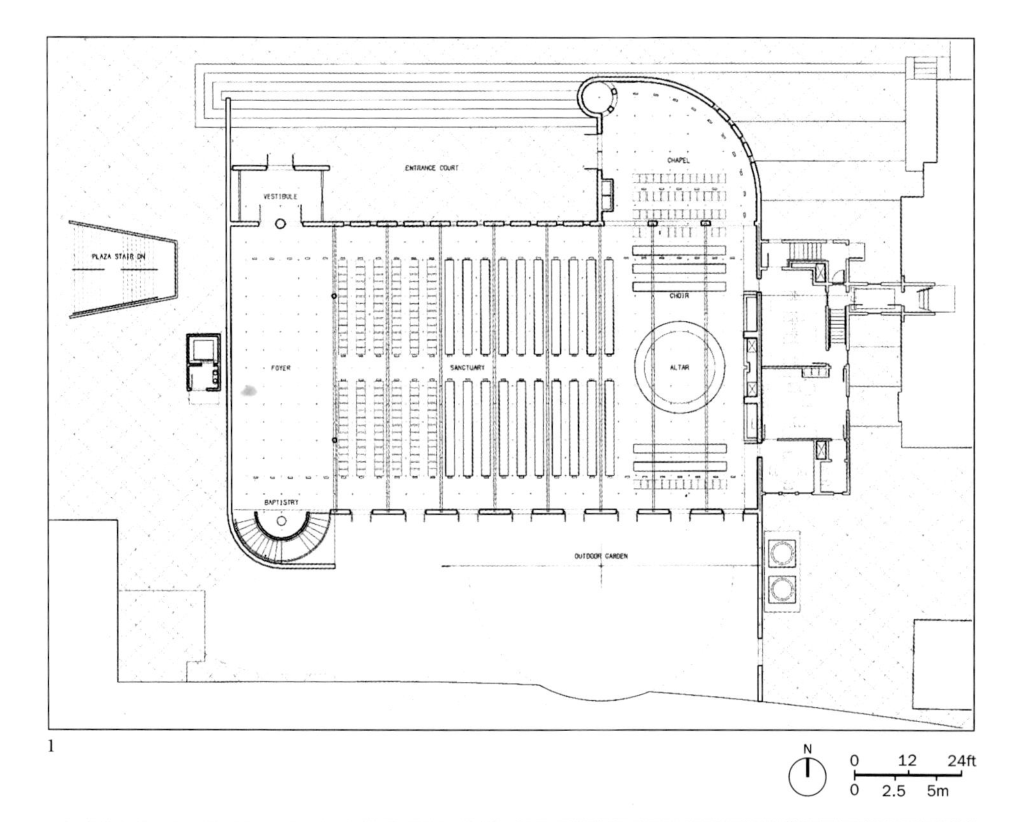

1

N
0 12 24ft
0 2.5 5m

2

1 Ground-floor plan
2 Street view looking east
3 West elevation: night view
4 Sanctuary altar looking north
5 Sanctuary view
6 Altar wall view

3

4

5

6

Islamic Cultural Center of New York

Design/Completion 1989/1991
New York, New York, USA
The Islamic Cultural Center of New York Foundation
21,000 square feet
Steel frame
Granite, glass, precast concrete and copper dome

The first religious center built for New York's Muslim community, the Center comprises a mosque, an assembly room, and a minaret. The building is rotated on the site to face Mecca. The facades are nine-square patterns that express the four large structural trusses — a configuration that supports the copper-clad, precast dome and provides a column-free worship hall.

Large glazed areas with fired ceramic surface decorations are enclosed within the metal trusses above the exterior granite walls. The dome rests on 16 articulated metal connections enclosed by clear glass, allowing a halo of light to permeate the prayer hall below. The interior is also illuminated by a ring of lights suspended on long bronze rods.

The 15-foot bronze doors open to reveal an abstract arch motif of layers of glass cut in rectilinear patterns; modern Kufic inscriptions are carved in granite above the doors. Opposite the entry portal is the mihrab, adorned with glass panels and a frieze of Koran verses.

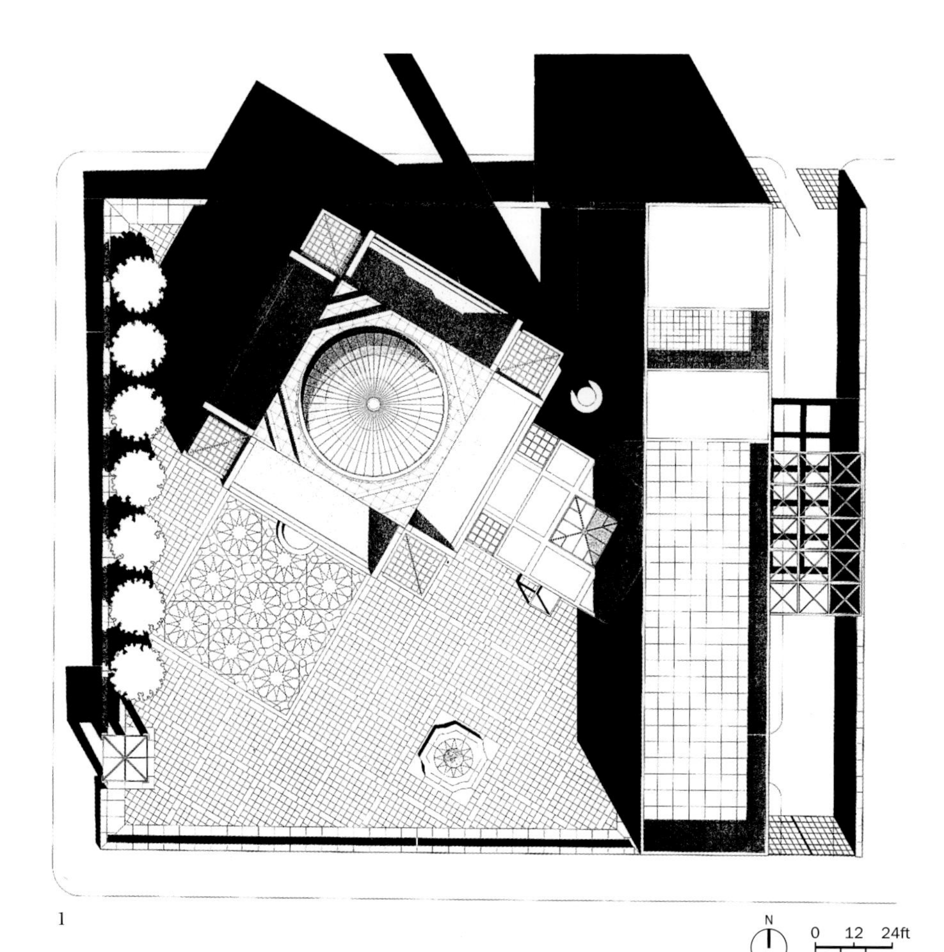

1

2

1 Site plan
2 Detail: corner skylight
3 General view

3

4 Interior of dome
5 Exterior detail
6 Mihrab
7 Main entrance
8 Prayer hall

4

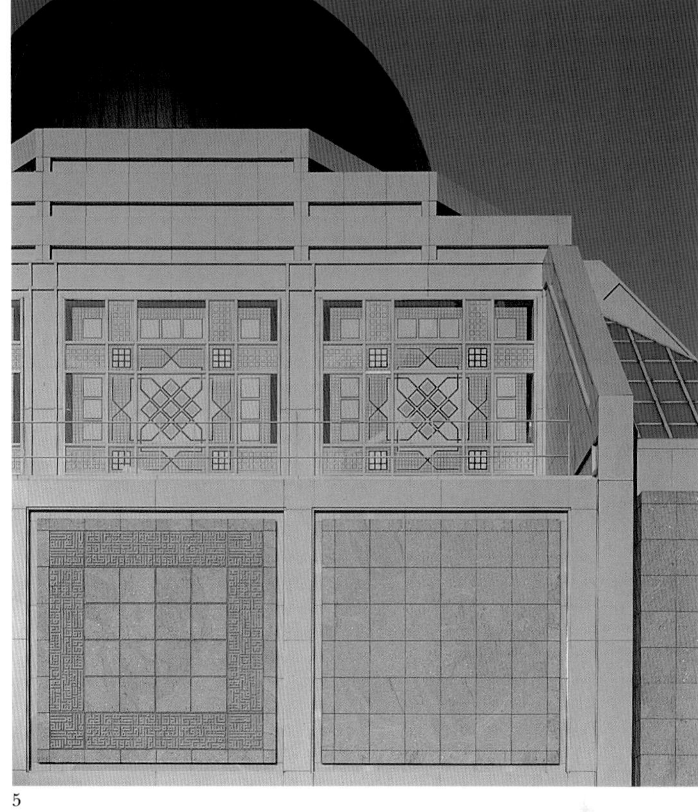
5

6

7

8

Aurora Municipal Justice Center

Design/Completion 1988/1990
Aurora, Colorado, USA
Aurora Municipal Building Corporation
208,000 square feet
Concrete and steel
Precast concrete

As the first increment of a civic building program, the Aurora Municipal Justice Center comprises a new courthouse, a new detention center, and an expanded police headquarters facility. The organizing axis runs from west to east, culminating in a large, domed rotunda which is the anchor of the complex.

In concert with local building practices, the design incorporates an architectural and structural precast system. The precast dome is the first of its kind constructed in the United States.

Departing from traditional court design, the courtroom configurations allow juries to face the witnesses. The design of one courtroom accommodates the remote arraignment of defendants and includes two high-security holding cells that are directly accessible from the detention center. Responding to the city's projected judicial requirements through the year 2010, the courthouse program focuses on flexibility for the future by planning for increases in the number of courtrooms and judges' chambers within the building envelope.

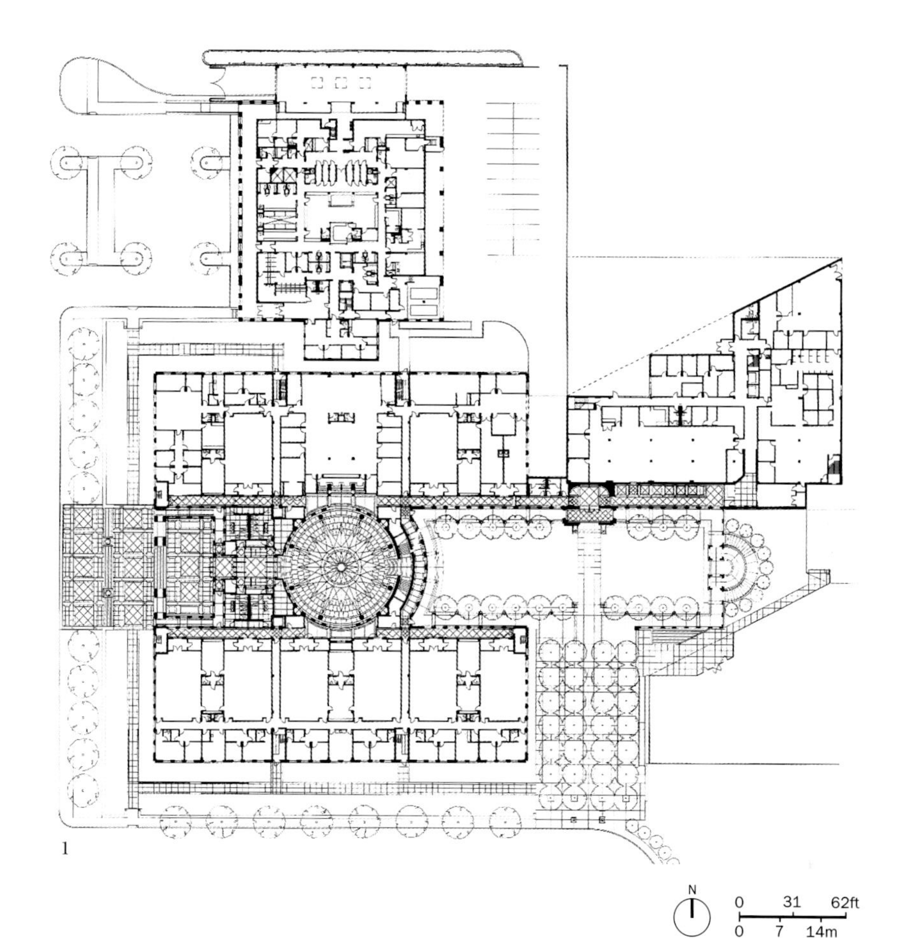

1

2

1 Ground-floor plan
2 View of inner court
3 General exterior view
4 View of exterior courtyard

3

4

5 Typical courtroom
6 Entry lobby showing rotunda

5

6

The Milstein Hospital Building

Design/Completion 1988/1990
New York, New York, USA
Columbia-Presbyterian Medical Center
839,000 square feet new construction; 497,000 square feet renovation
Steel structure
Veneer masonry

As the centerpiece of the Presbyterian Hospital modernization and consolidation, the Milstein Hospital Building houses 745 beds, ancillary diagnostic and treatment areas, and 22 new operating rooms.

The grid established by brick masonry with stacked bond joints contrasts with the existing hospital buildings; however, the beige-hued masonry relates the building to the larger context. The long facade of the main building is deliberately emphasized to reinforce the hospital's prominence as a major city institution—a teaching hospital known for its many areas of specialization. The building fronts on the Hudson River and is visible from both Riverside Drive and the Henry Hudson Parkway.

A combination galleria and concourse, the "Energy Court," serves as the major east–west circulation link between the existing complex and the building, creating a new front door on Broadway as well as providing access to all existing buildings. Public amenities such as shops, lounges, and recreational and educational facilities are located within that link.

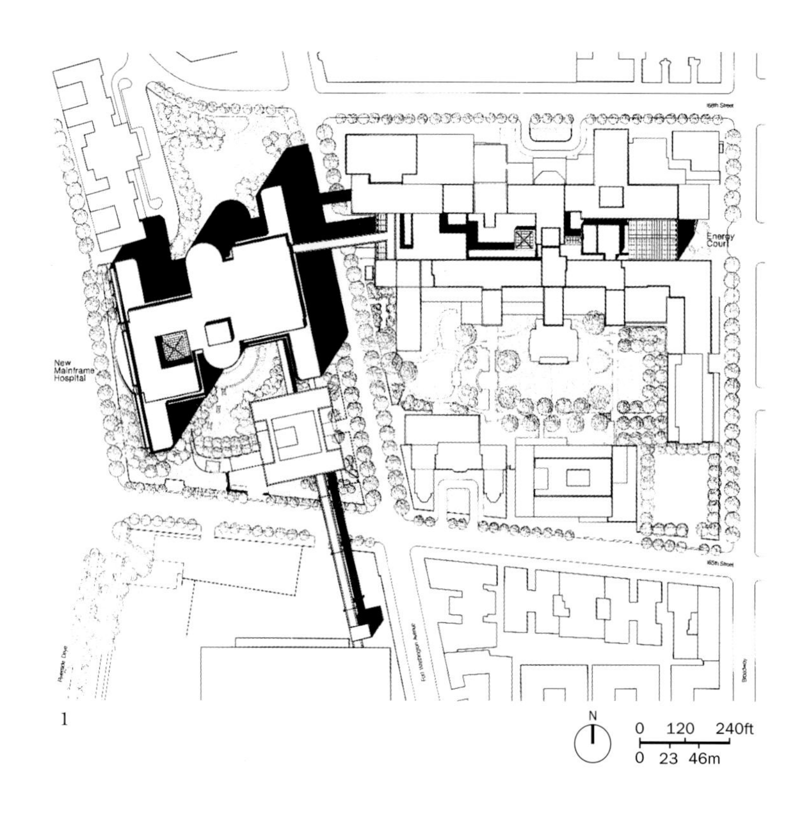

1

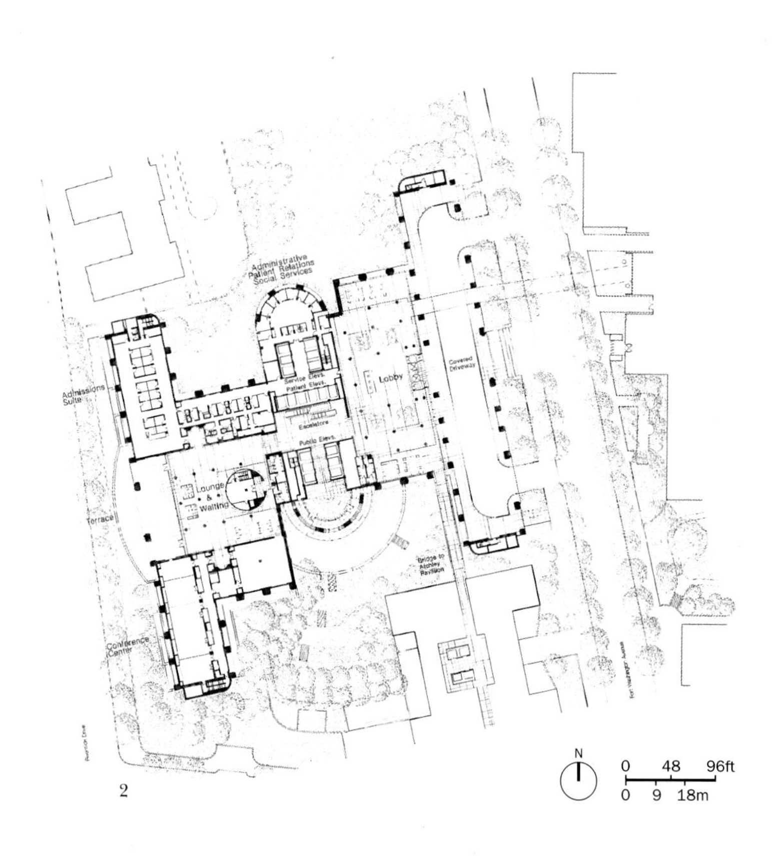

2

1 Site plan
2 Ground-floor plan
3 View from garden courtyard
4 View of connecting skybridges
5 View from river

3

4

5

6 McKeen Pavilion: atrium
7 McKeen Pavilion: patient-care suite
8 Ambulatory surgery recovery
9 Intensive care unit

6

7

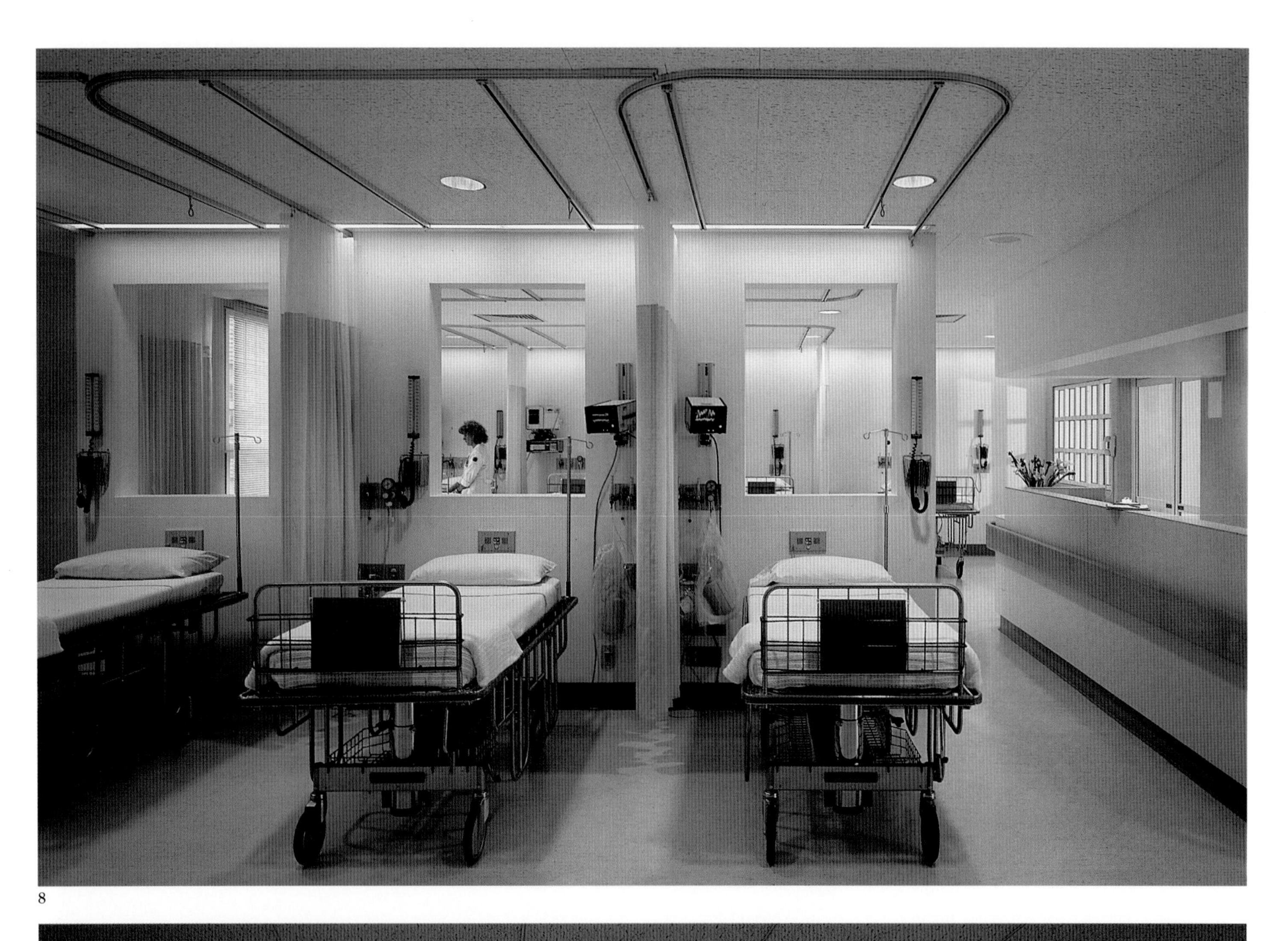

8

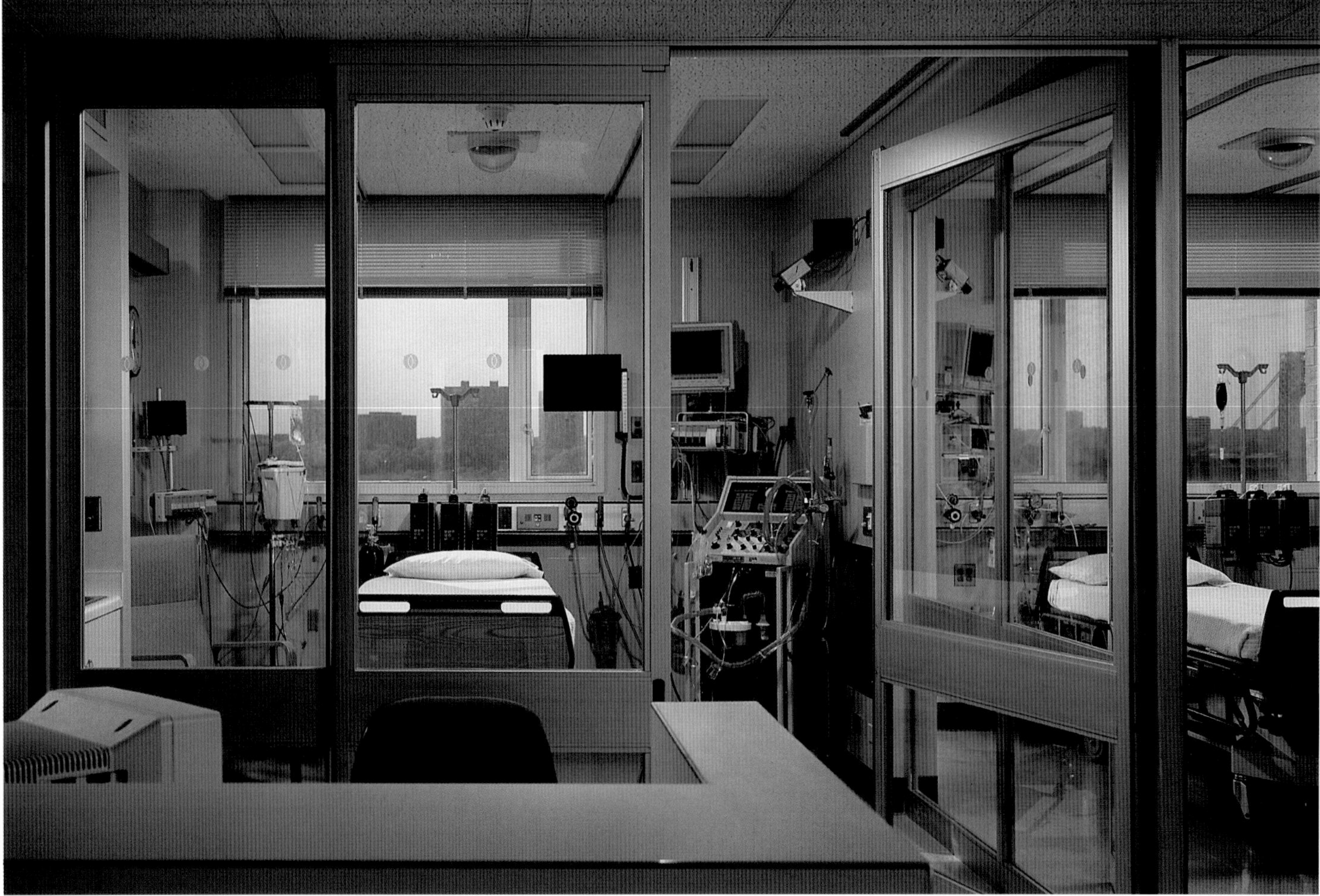

9

Yongtai New Town

Design/Completion 1994
Guangzhou, China
New World Properties Ltd

The plan creates a lively development that will activate the northern suburbs of the Guangzhou metropolitan area. The development complements neighboring cultural, recreational and commercial projects contained within the recently formed South Lake National Tourist and Vacation District of Guangzhou. These include the New Orient Theme Park; Asia Movie City; Commercial and Trade City; the Convention Center; the International Exposition Center; the Mountaintop Scenic, Tourist and Vacation Center; the zoo; and the Baiyue Cultural Village and International Park.

The conceptual plan gives the development program with a sequence of buildings, spaces and public amenities that will result in a harmonious environment. A hierarchy of streets, pedestrian walkways, and open-space amenities provides variety throughout the plan. A naturalistic, meandering park along the banks of a winding stream creates critical links between the urban core, residential areas, and the activity of the town center's piazza and formal lake.

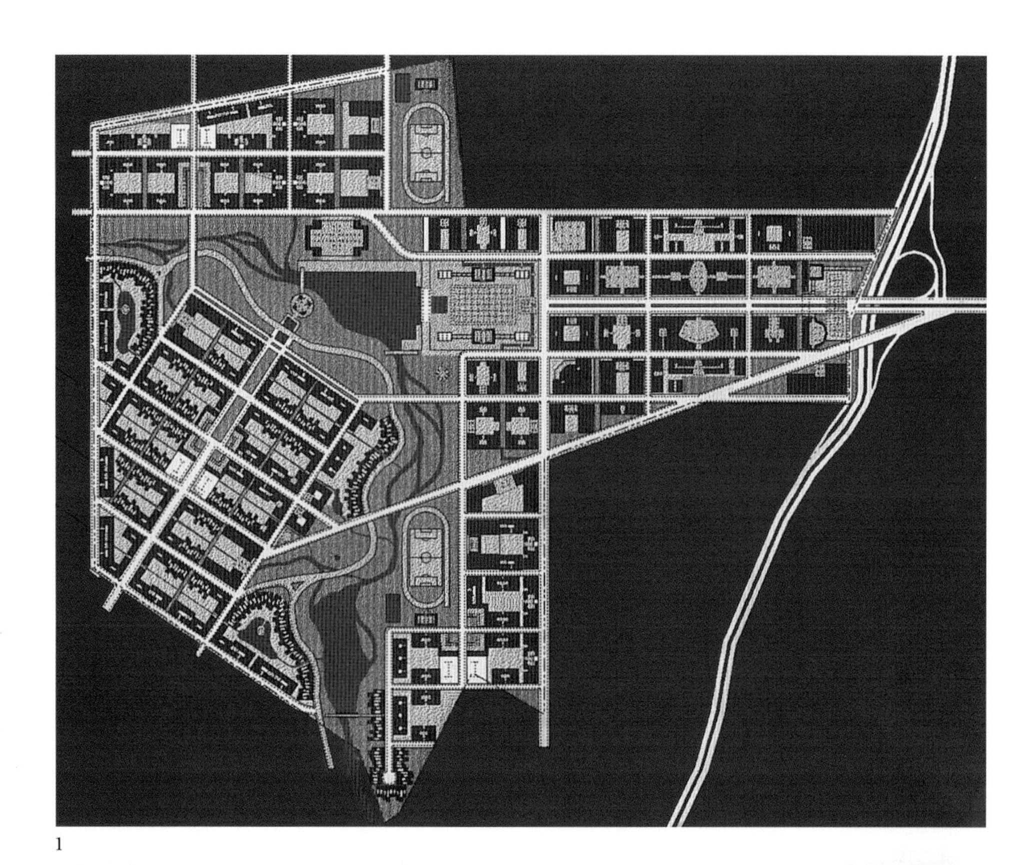

1

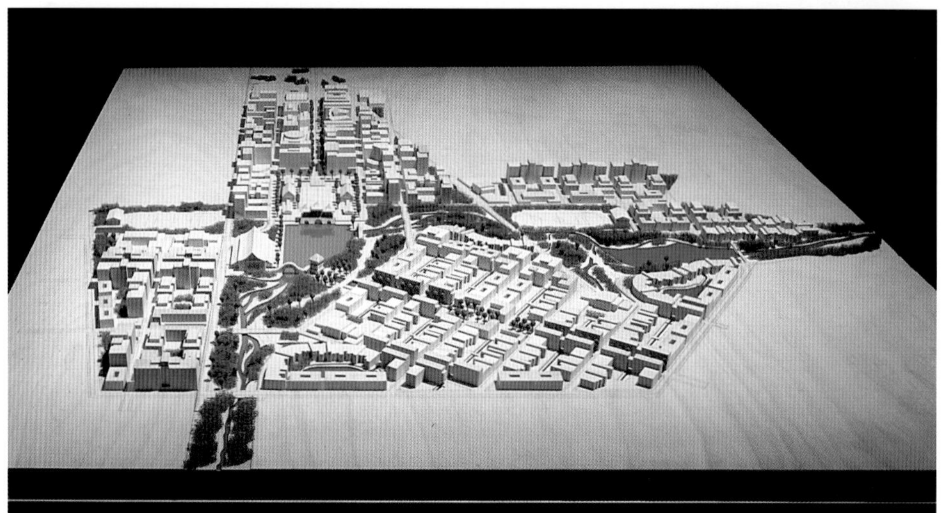

2

1 Master plan
2 Aerial view of model
3 Conceptual organization diagram
4 Vehicular and pedestrian circulation diagram

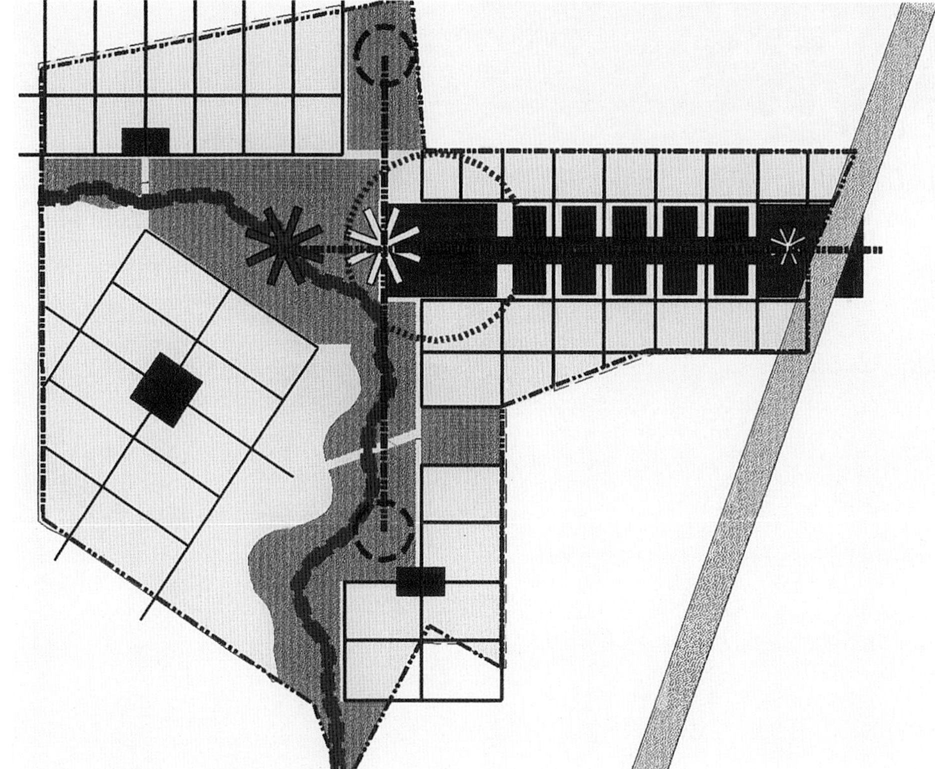

3

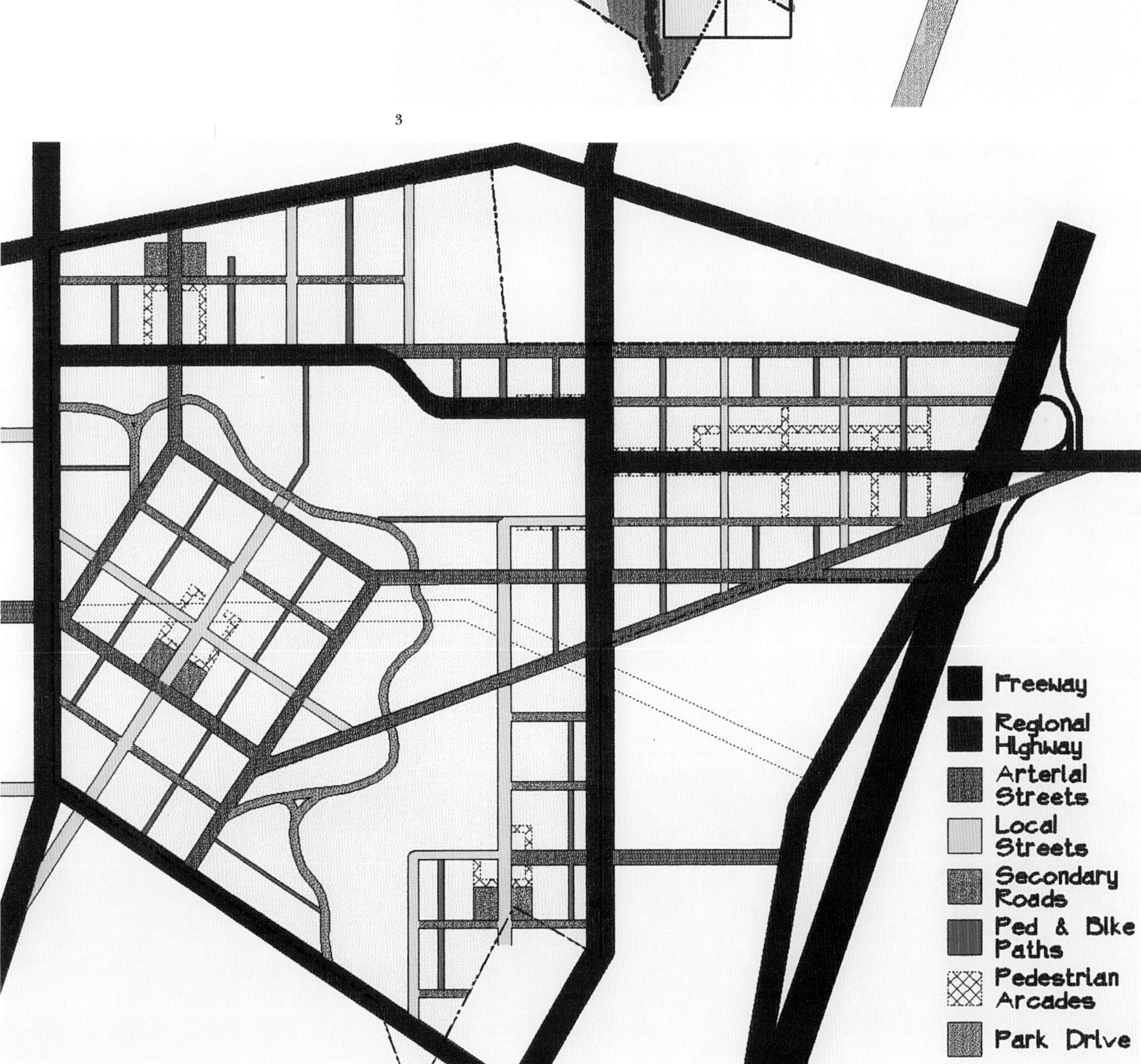

4

Transitional Housing for the Homeless

Design/Completion 1987/1993
New York, New York, USA
City of New York Department of General Services
700,000 square feet
Concrete; CMU wall and precast plank
Brick veneer

As models for a public program providing housing for homeless singles and families, both schemes offer amenities such as medical care and a range of social services.

The families building houses approximately 100 families. Organized around a central entrance pavilion with social services and communal areas, the residential wings contain both single and double living units. The brick buildings are characterized by painted aluminum windows and a hipped roof on the entrance pavilion, which suggest traditional house forms. Outdoor space and playgrounds are part of the program.

The singles facility houses approximately 185 residents in "clusters" stacked in an interlocking manner throughout the building. Common facilities and dining rooms are centrally located and link different site configurations.

1

2

1 Families: typical bedroom
2 Families: typical kitchen and dining
3 Families: general view
4 Families: playground

3

4

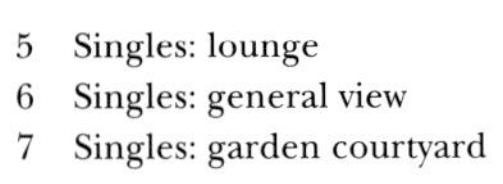
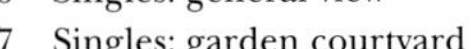

5 Singles: lounge
6 Singles: general view
7 Singles: garden courtyard

5

6

7

Kirchsteigfeld

Design/Completion 1993/1995
Potsdam, Germany
GROTH + GRAALFS GmbH
200,000 square feet
Reinforced concrete frame
Stone, render

The building is sited on the main square of a new mixed-use development in the south-eastern suburb of Potsdam, near Berlin. Rob Krier, who designed the original master plan, envisaged designs by a number of different architects for each portion of the development.
SOM's design incorporates office and residential components, and includes an underground parking garage and retail facilities for building residents, office workers, and neighborhood residents.

The design of the building complies with the established master plan guidelines. Features such as overall heights and roof slopes respond to the scale and architectural features of adjacent buildings. The simplicity of these building forms requires the use of a reduced palette of materials dominated by stucco finishes. A prominent tower completes the view at the end of the main south–north axis.

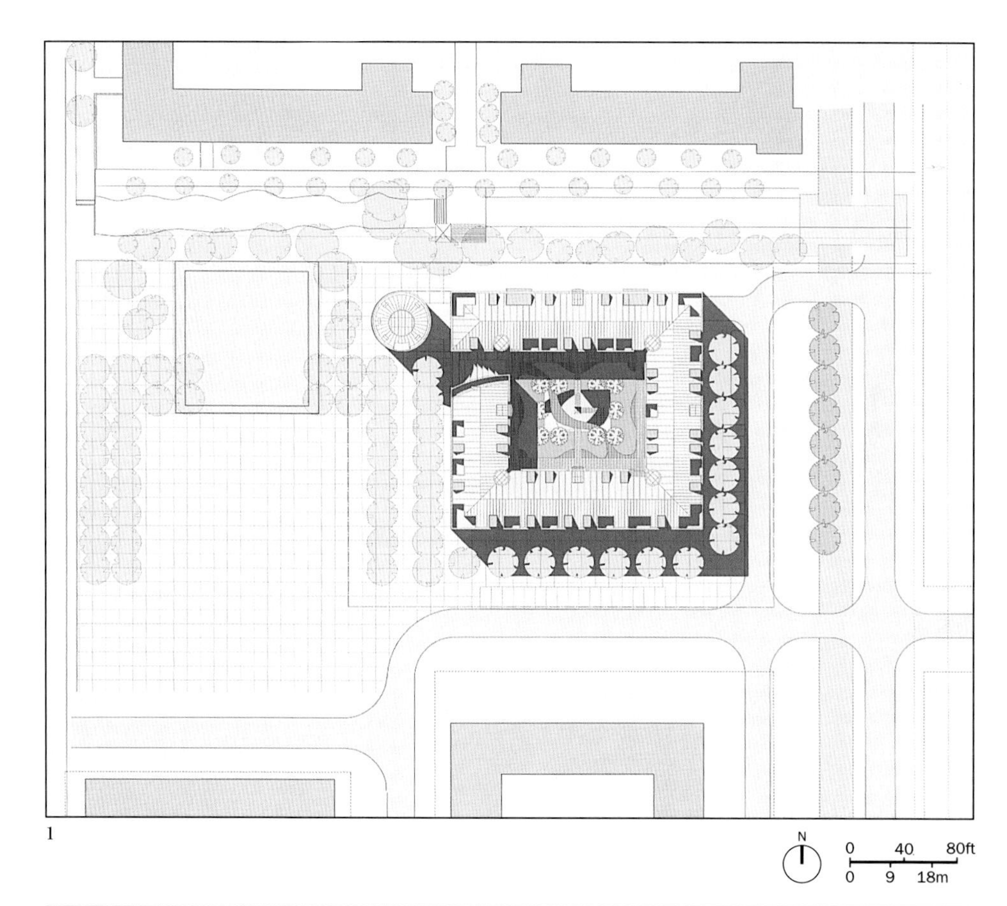

1

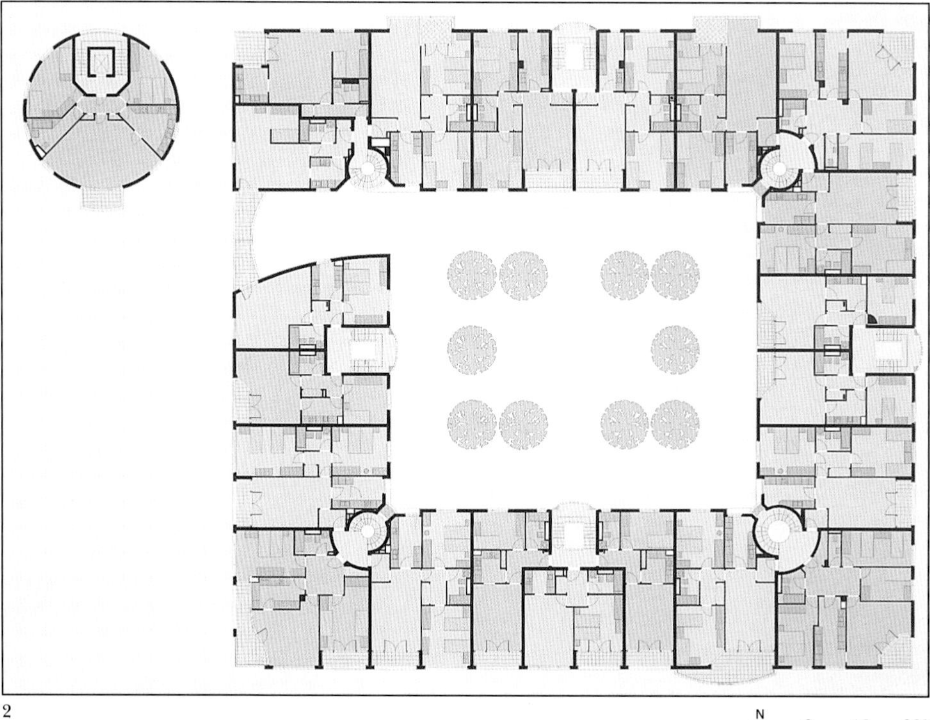

2

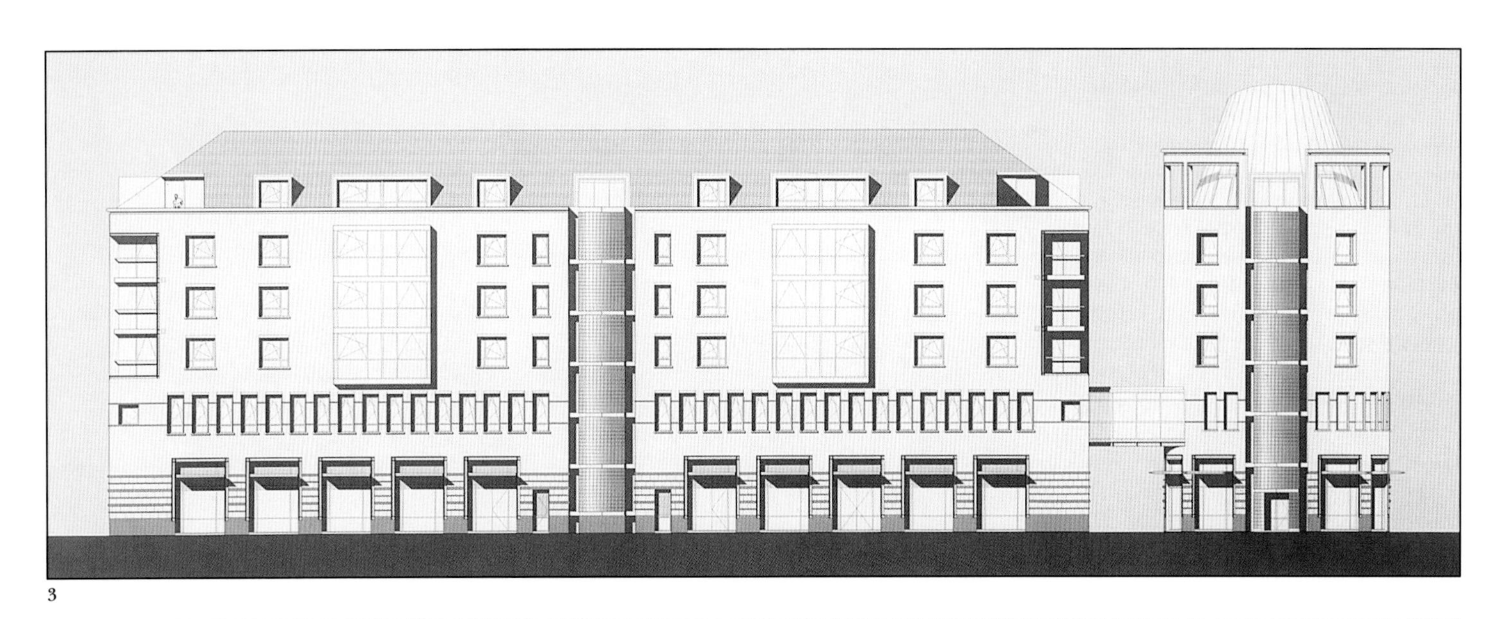

3

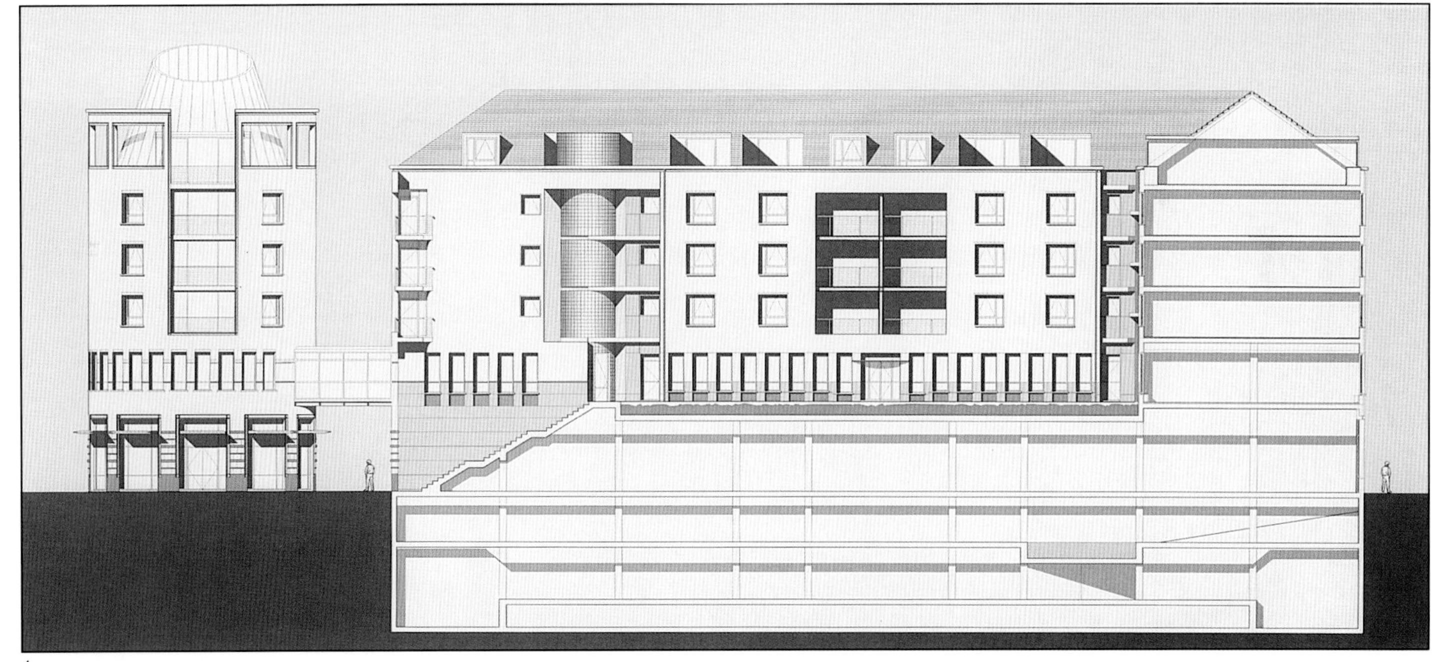

4

1 Overall site plan
2 Plan of typical residential floor
3 North elevation to linear landscaped park
4 Section through interior courtyard

Northwest Frontier Province Agricultural University

Design/Completion 1985/1994
Peshawar, Pakistan
US Department of State/Agency for International Development, Mission to Pakistan
465,000 square feet (site area: 430 acres)
Reinforced concrete frame
Reinforced brick infill, marble veneer, colorcrete, wood-frame windows

The project master plan recommends the expansion of an existing campus to accommodate a doubled student population and admission of female students. The first phase includes 85 buildings, incorporating new teaching and research facilities, library/media and computer centers, administrative offices, an auditorium, housing for 475 students and 50 faculty members with families, and new infrastructure providing complete electric, gas, water and sewer utility distribution, roads, bridges, playing fields and landscaping.

The design solution builds upon existing renovated facilities, creating a central academic campus surrounded by clusters of student housing, activity fields, and support facilities. Faculty housing is located across the river that bisects the site, along with plant and animal research centers adjacent to experimental crop fields.

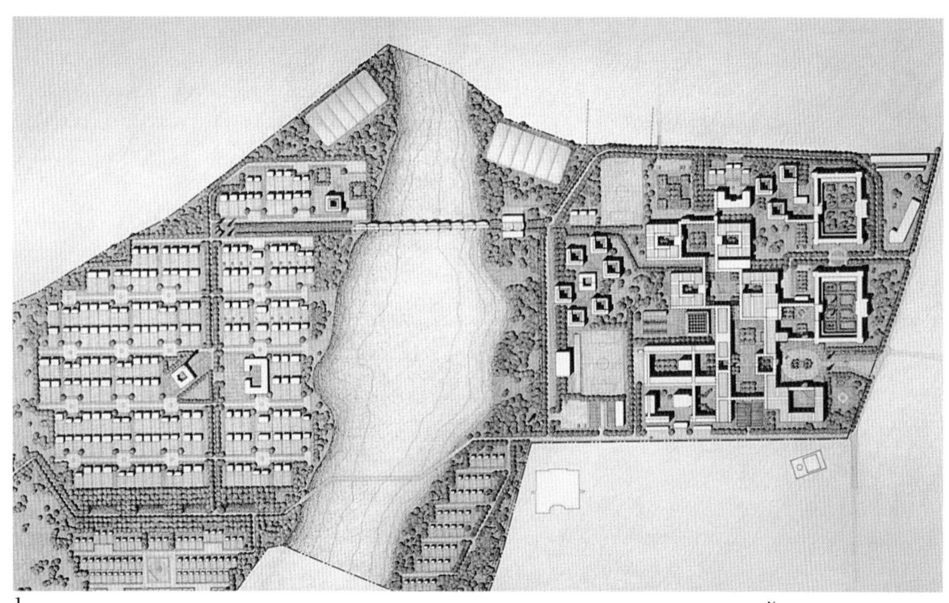

1

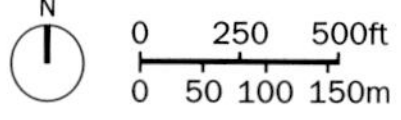

2

1 Illustrative master plan
2 Library reading room
3 Academic courtyard arcade

3

Northeast Corridor Improvement Project

Design/Completion 1977/1984
Washington, DC to Boston, Massachusetts, USA

The project is a $2,500,000,000 program of the US Department of Transportation to improve intercity rail passenger services along the 456-mile corridor linking Boston, New York City, and Washington, DC. As part of the multidisciplinary team, SOM established architectural design standards for the entire system and realized improvements to 15 major passenger stations along the corridor. Stations ranging from suburban facilities with commuter and intercity passenger services to major urban terminals of outstanding architectural or historic significance are included in the project, and include Boston's South Station, Providence's Intercity Rail Station, Philadelphia's 30th Street Station, Wilmington Station, Baltimore's Penn Station, and Washington's Union Station.

The program encompassed the preparation of detailed designs for restoration, renovation or new construction at each station; preparation of environmental impact assessment documents; and participation in securing the government approvals required to implement the station plans.

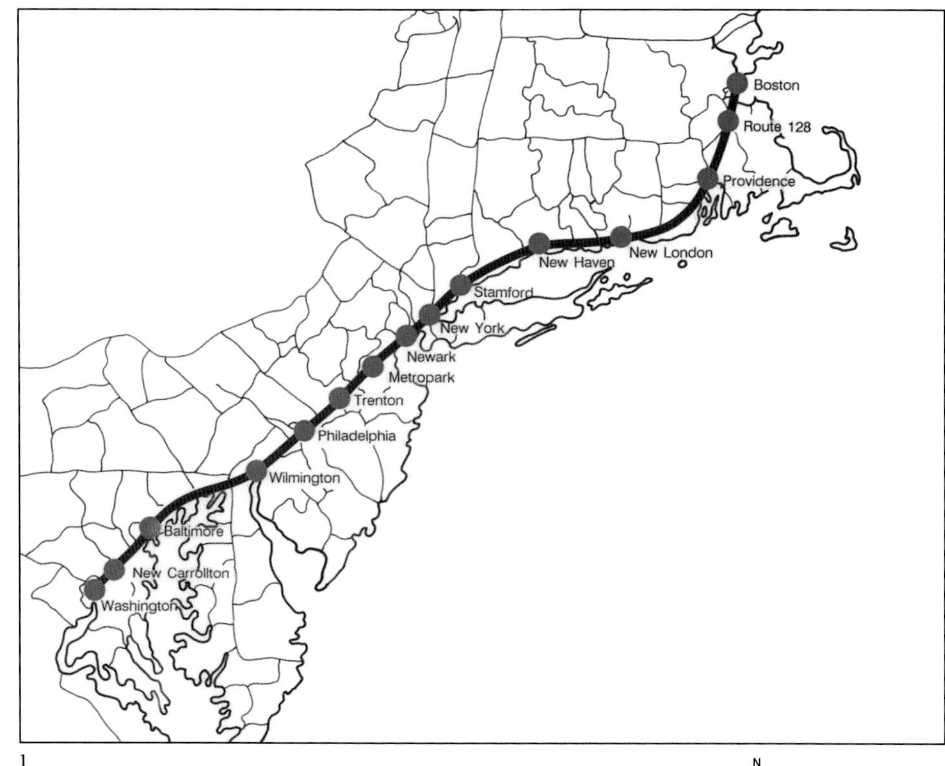

1

1 Map of the Northeast Corridor
2 Providence Intercity Rail Station: exterior
3 Wilmington Station, Wilmington, Delaware

2

3

4

4 Baltimore Penn Station: interior
5 Baltimore Penn Station: exterior
6 Baltimore Penn Station: wall detail
7 Baltimore Penn Station: lighting detail

5

6

7

Dulles International Airport Main Terminal Expansion

Design/Completion 1989/1997
Chantilly, Virginia, USA
Washington Metro Airports Authority
1,026,000 square feet
Concrete structure, catenary roof
Glass curtain wall, concrete

As the earliest realization of the separation of landside and airside components, Dulles became an architectural and functional model for much subsequent airport design after its completion in 1962.

The current expansion of the landmark Main Terminal building, designed by Eero Saarinen in the late 1950s, increases the annual passenger handling capacity from 12 million to 50 million passengers. The catenary roof structure and podium are be replicated to the east and west, housing ticketing, check-in, baggage claim and support functions. The expansion of below-ground facilities for baggage handling, security, and passenger transport to airside under the apron minimize visual impact, as well as the disruption of aircraft circulation. The project also includes the demolition and reconstruction of enplanement and deplanement ramps and the overall renovation of the existing terminal. The firm also conducted the expansion master plan review and designed the 1991 International Arrivals Building.

1

2

1 Section through main terminal and security mezzanine
2 Perspective view from departures roadway
3 Computer ray-trace of security mezzanine
4 Computer ray-trace of people-mover platform

3

4

Logan Airport Modernization Program

Design/Completion 1993
Boston, Massachusetts, USA
MassPort
Concept plan

The commission for this project was won in a limited competition. In a parallel effort with the concept planning phase for the one billion square foot expansion of the airport, the program establishes comprehensive design guidelines, and sets the architectural imagery and vocabulary for the airport.

Airport components detailed in the design guidelines include a new Terminal A, the expansion of the existing Terminal E, a parking garage, a hotel, a people-mover system, airport roadways, and a complete landscape program. The guidelines identify design elements that range from the forms of interior and exterior architectural expression, to the relationships between new and existing facilities, to the type and quality of finishes and the selection of materials and colors.

1

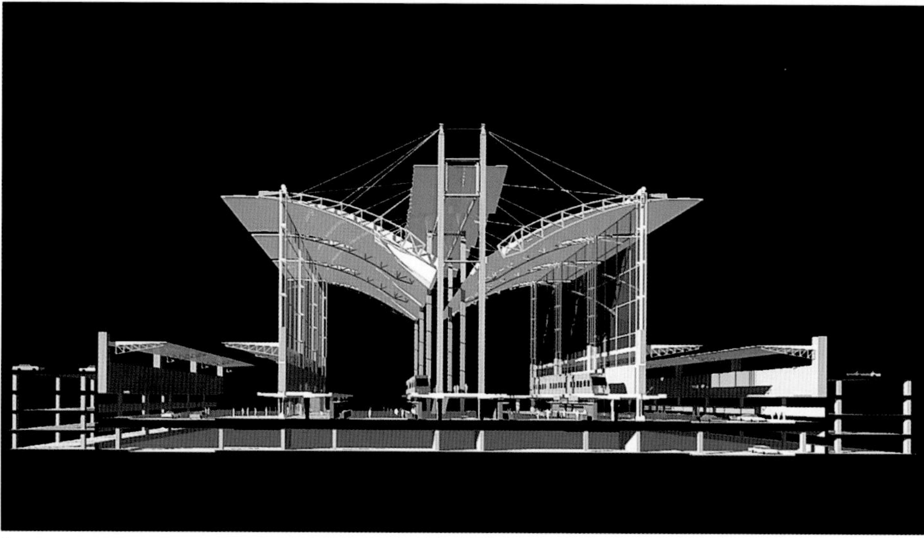

2

1 Aerial perspective of the site
2 Computer ray-trace of central terminal
3 Computer ray-trace of departures concourse
4 Computer section of central terminal

Terminal B
Air Transportation Center
Baggage Claim
Restrooms
Elevator
Telephone
Gates A1-A20
Passenger Service
Information
Gate
A21
10:22
Gate
A21
Gate
A20
Gate
A22
AA 822
A22

3

4

International Terminal, San Francisco International Airport

Design/Completion 1994/1997
San Francisco, California, USA
San Francisco Airport Commission
1,600,000 square feet
Steel frame structure
Glass and metal curtain wall

The centerpiece of the airport's current expansion program, the new International Terminal includes all facilities for international arrivals and departures, including duty-free shops, restaurants, and airline lounges. The building also houses airport tenant offices and serves as a central terminal for the airport's new light rail system.

The wing-like roof over the departure hall is a principal feature of the design. The roof is a double cantilever that follows the moment diagram of the structure. Daylight gives it a floating appearance, while the clear and patterned glass of the curtain wall at the entrance gives the interior a transparent and luminous quality.

The joint venture proposal of SOM, Del Campo & Maru, and Michael Willis & Associates was selected after winning an invited design competition sponsored by the airport.

1

2

1 View west from existing airport
2 Interior view, the International Terminal "Great Hall"
3 Montage: computer study of roof structure and inset site plan
4 Model study: International Terminal from approach road

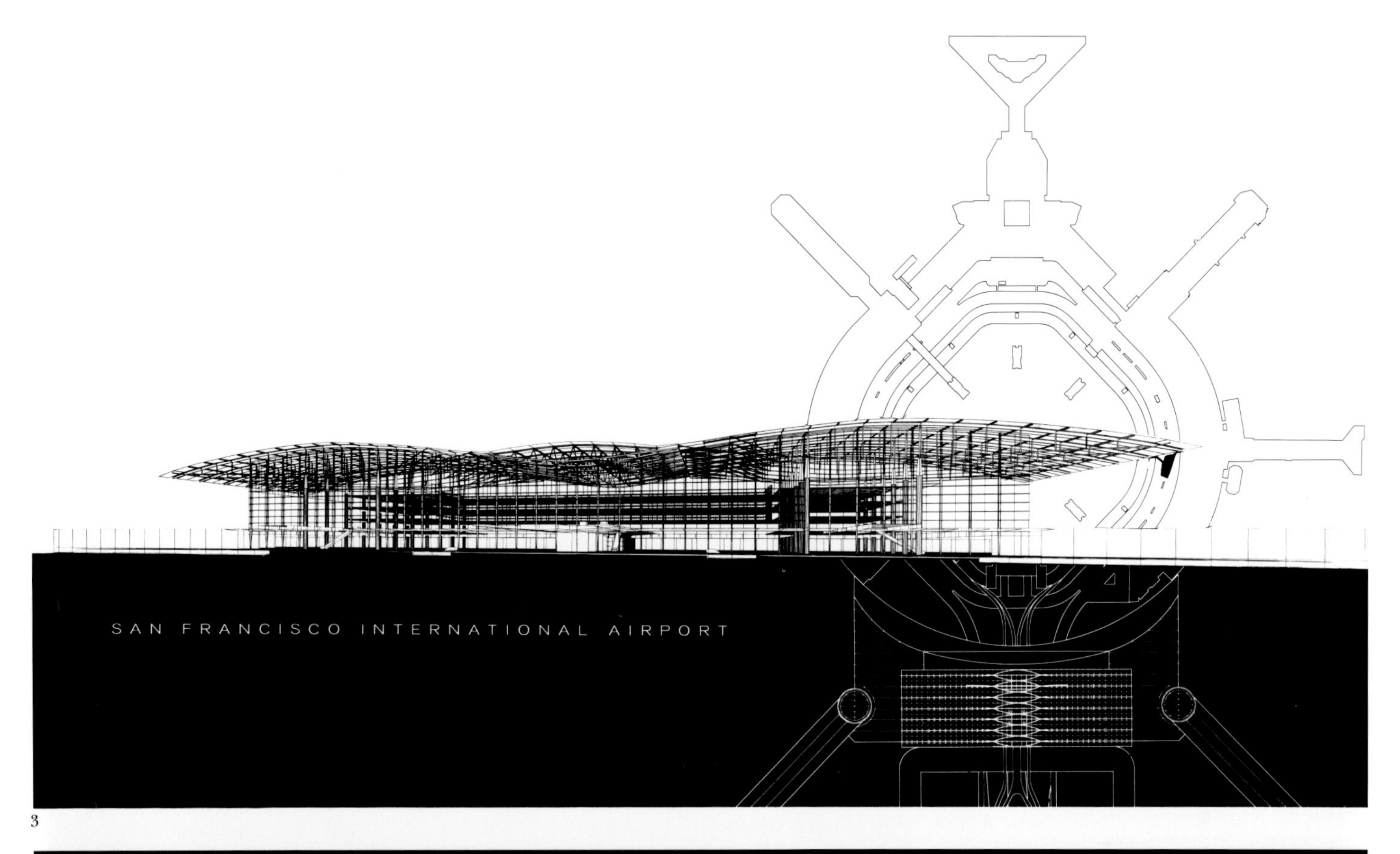
SAN FRANCISCO INTERNATIONAL AIRPORT
3

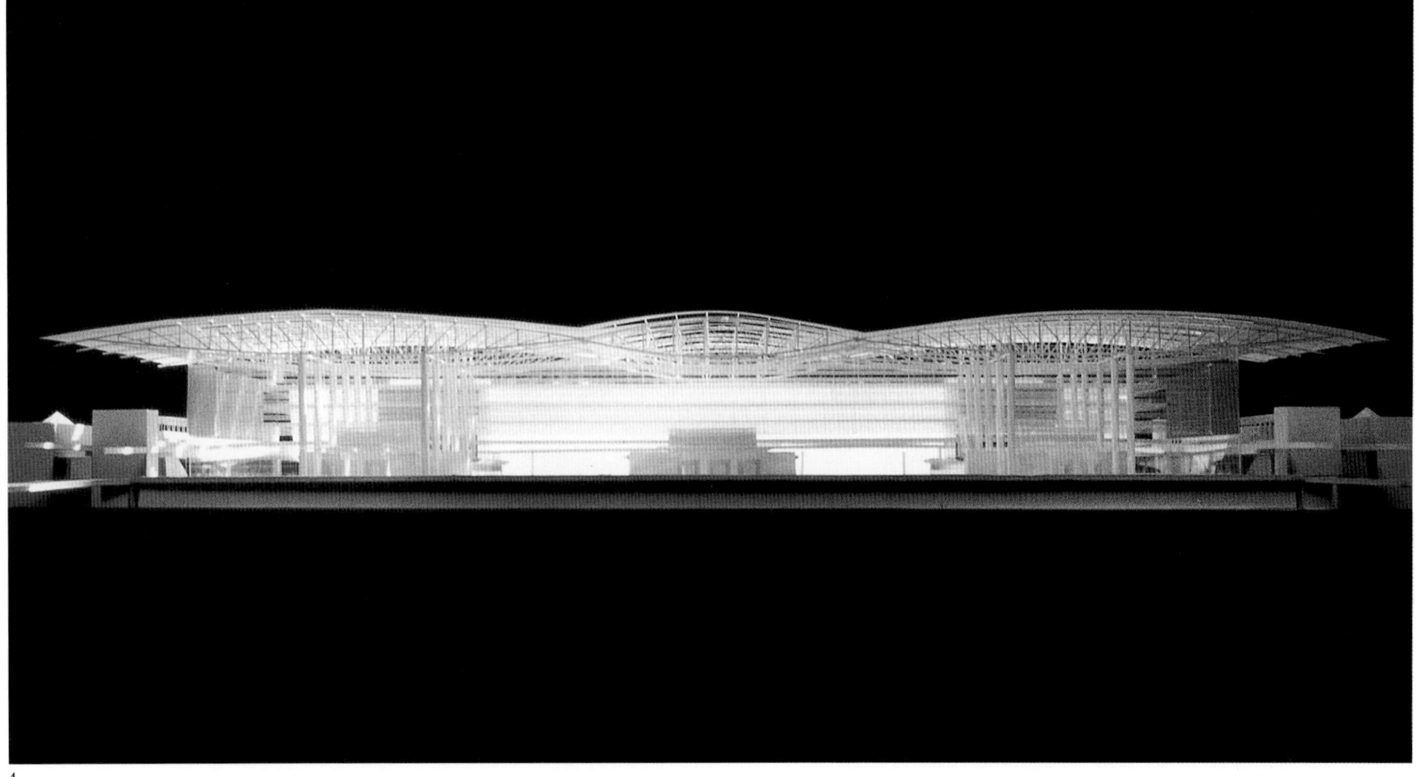
4

New Seoul Metropolitan Airport Competition

Competition 1992
Seoul, South Korea
Korea Airports Authority
Steel roof over tubular steel truss structure
Glass and aluminum panel wall system

The phased-construction airport is technologically advanced, operationally efficient, passenger-friendly, cost-effective, and sets a new standard for international airports in Asia. The design blends advanced technology with elements traditional in Korean architecture—harmony with nature, order and spatial hierarchy, and the use of natural light.

The cohesiveness of the complex, which includes the terminal, control tower, airport access roads, rapid transit system, and International Business Center (IBC), is critical to its success. The IBC concept, one of the most innovative strategies of the overall airport master plan, incorporates numerous passenger amenities including convention center/exhibition buildings, retail galleria, a hotel, conference halls, together with office space and research and development facilities.

The Phase I design consolidates the airside terminal and transportation/processing functions into one building, and includes the rapid transit system and bus station. The terminal's curved form facilitates future expansion without disruption of airport operations.

1

1 View of main terminal
2 Airside view of project
3 Main departure hall

2

3

KAL Operations Center, Kimpo International Airport

Design/Completion 1991/1995
Seoul, South Korea
Korean Air Lines
1,300,000 square feet
Steel truss

The multi-purpose building combines hangar operations with facilities for airline employees, including executive suites, a healthcare center, flight crew lounges, and support staff areas. Primary programmatic elements are the wide-body aircraft hangar (295 feet x 590 feet clear span space) and the annex building, which wraps around the hangar in a U-shaped configuration. Each level of the facility serves a specific function, with employees from upper and lower levels meeting at a common floor designed for centralized events.

Innovative structural and design solutions required by the project include the development of a new structural roof framing system. The double-arch steel truss roof is supported principally on three box columns, separated by an expansion joint from the annex building. However, three major columns are connected to the annex building at several levels. This system employs standard building and roofing materials in an entirely new way.

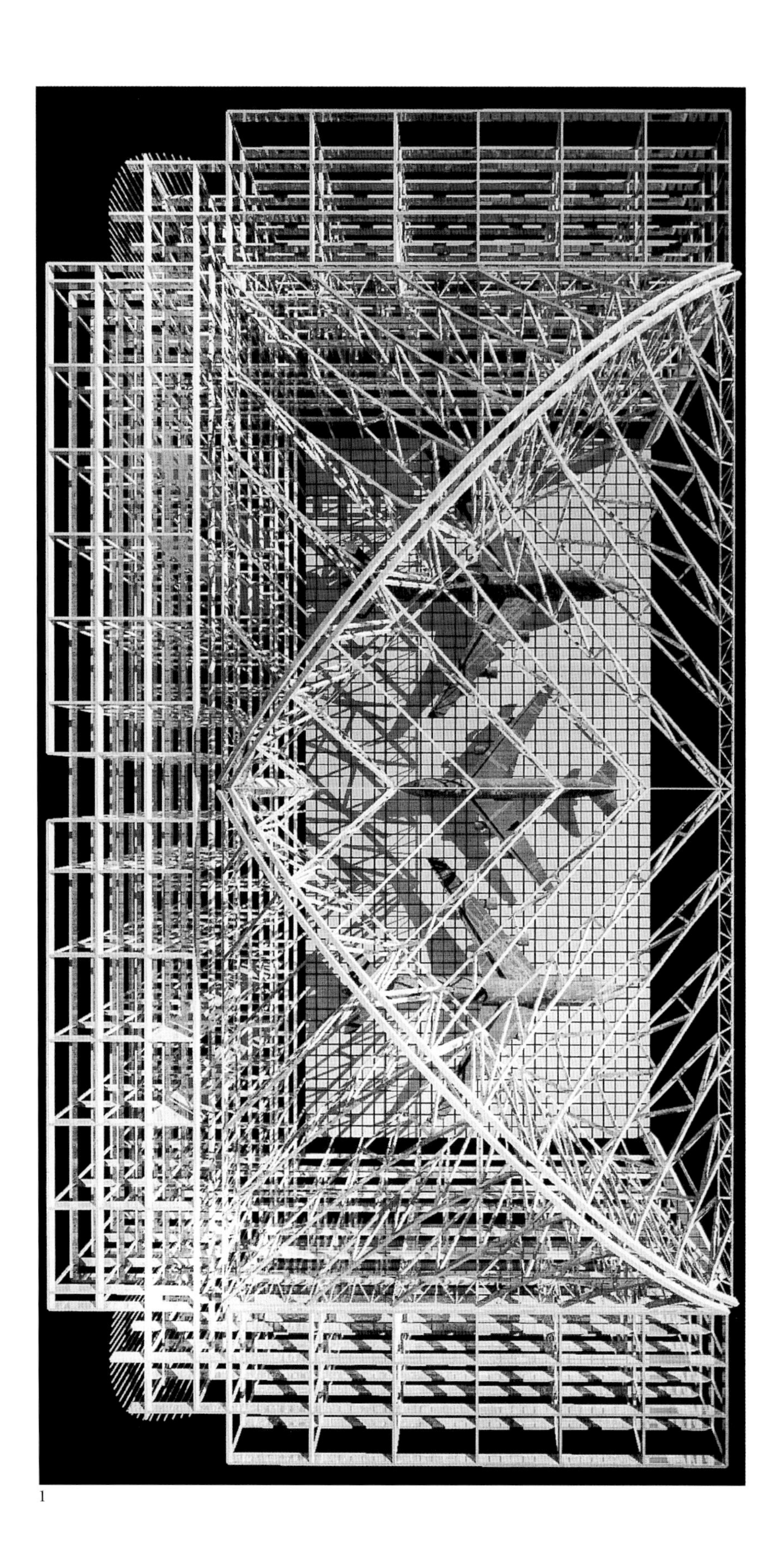

1

1 Computer rendering: view of roof structure from above
2 Computer rendering: interior view of hangar
3 Model: airside view of the new facility

2

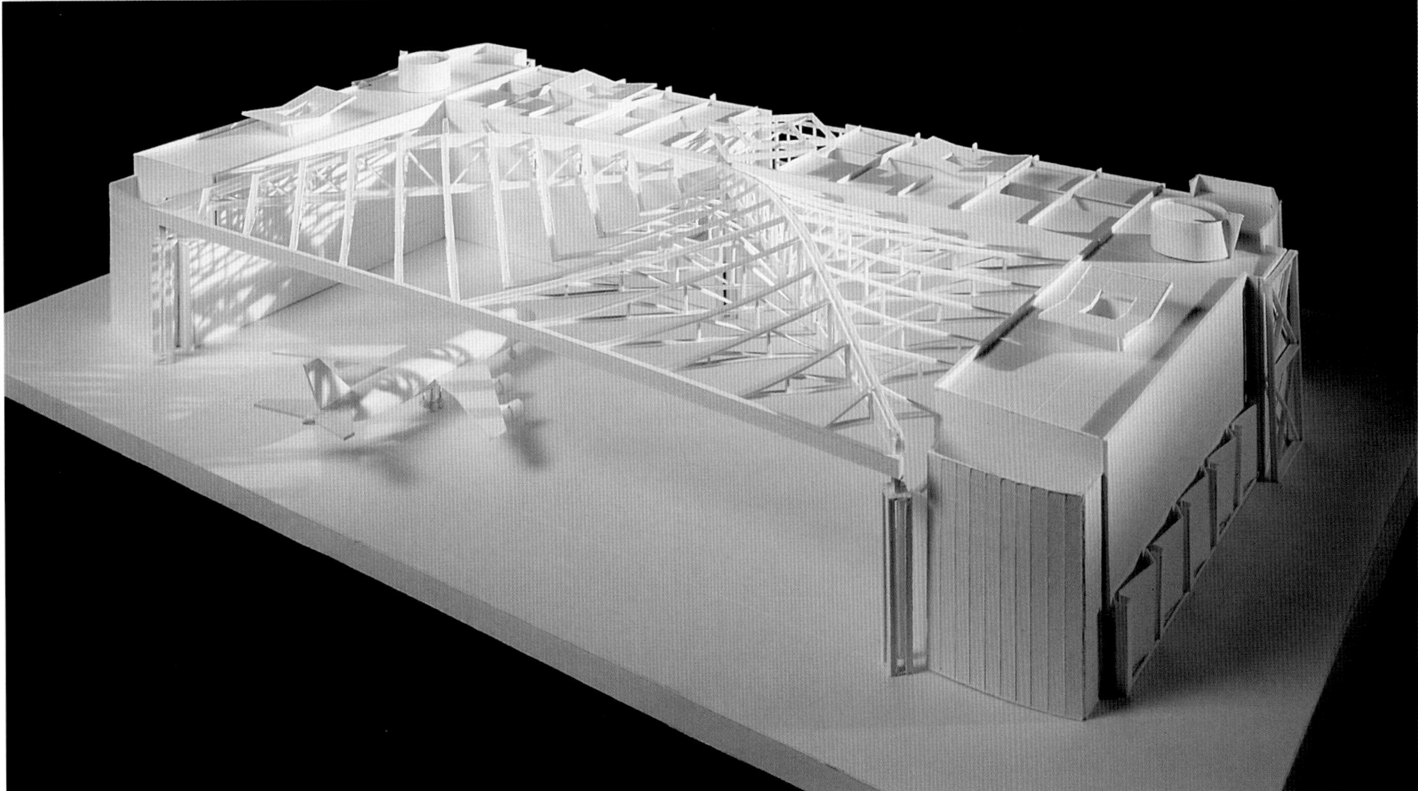

3

Tribeca Bridge

Design/Completion 1991/1994
New York, New York, USA
Battery Park City Authority
Span: 299 feet
Steel structure with bowstring truss
Glass cladding system

The pedestrian bridge provides a safe, grade-separated crossing for students and area residents over the busy West Street/ Route 9A corridor, connecting the new Stuyvesant High School to the public open spaces of Battery Park City.

Bowstring trusses on either side of the glass and wire mesh walkway are offset from each other in response to the surrounding street grid. At the center of the span, the painted white trusses form an arch, enclosing the horizontal glass and metal walkway that is suspended within the structure.

The elevator towers mark the east and west ends of the bridge, and are ornamented by stainless steel details at the tower tops and entry canopies. The eastern stair, constructed of cast-in-place concrete, curves around the elevator shaft in the tradition of a grand staircase. The stairs at the western terminus of the bridge provide egress for pedestrians not entering the school.

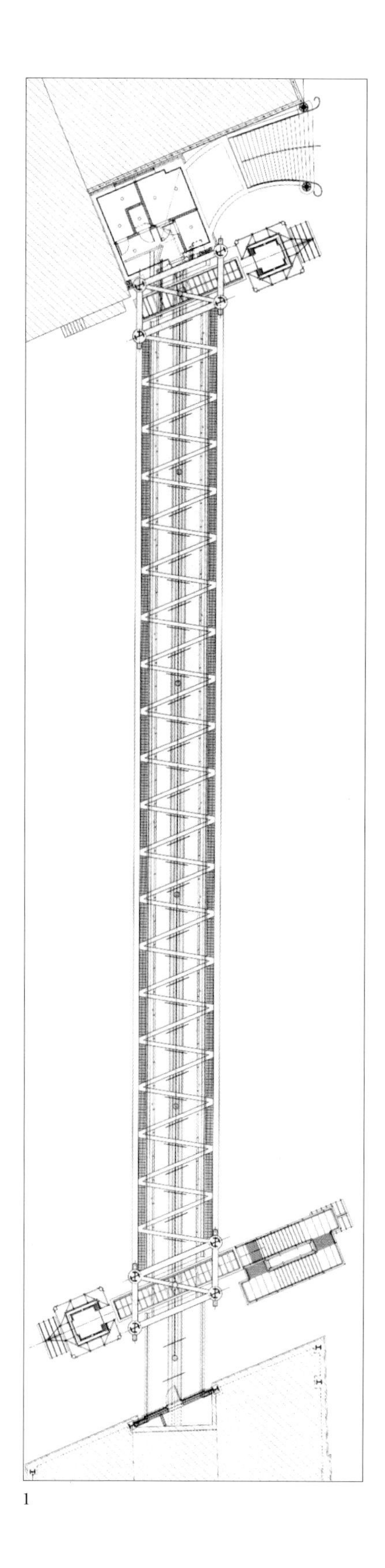

1

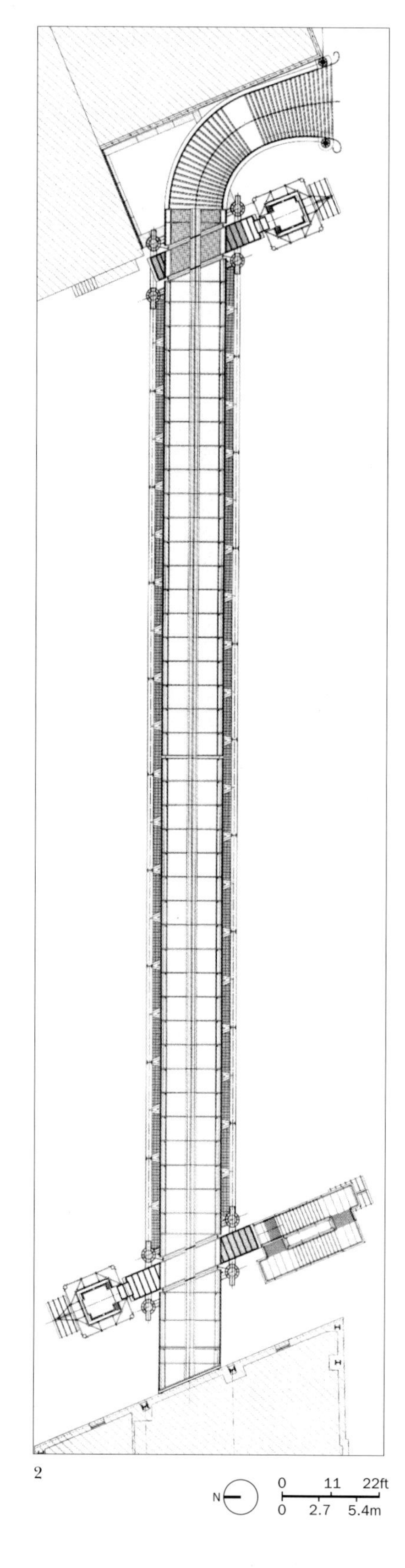

2

N

0 11 22ft
0 2.7 5.4m

1 Reflected ceiling plan
2 Roof plan
3 Elevator tower (clockwise from top left: elevation, section, roof plan, ground-floor plan)

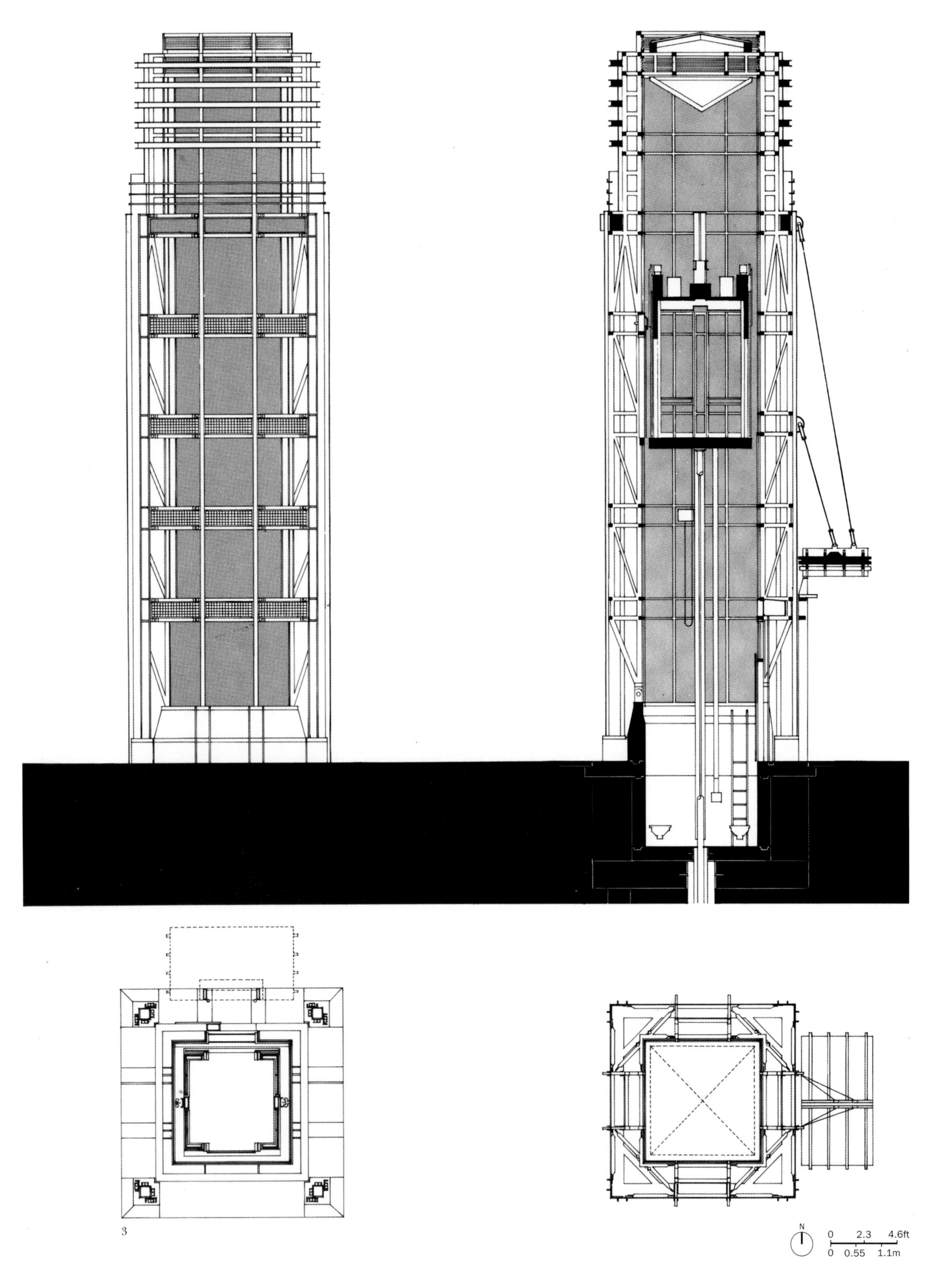
3
N
0 2.3 4.6ft
0 0.55 1.1m

Morrow Dam

Design/Completion 1985/1986
Kalamazoo County, Michigan, USA
STS Consultants Ltd
Steel frame
Concrete, glass block, glass

The project is a hydroelectric power plant housing machinery and related dynamic operations. A prismatic triangular structure continues the rhythm of the existing spillway expressing the strength of the power housed within. The design uses the triangle as a basic form, the steel bent as a modulator, and the green glass sheathing as a reflector of the sky, water, and landscape.

For general maintenance, removal, and replacement of the hydroelectric machinery within the plant, a crane was incorporated into the structure. The crane can travel freely longitudinally and transversely throughout the space. A computerized system, operated from the control room at grade level, monitors the hydroelectric machinery. The dam also acts as a passive solar collector.

1

1 Interior view from generator level
2 Exterior view looking north-east
3 East elevation

2

3

Commonwealth Edison Company, East Lake Transmission and Distribution Center

Design/Completion 1991/1998
Chicago, Illinois, USA
Commonwealth Edison Company
100,000 square feet
Reinforced concrete structural frame
Granite, limestone, glass, stainless steel, copper, painted aluminum

The project creates a pedestrian focus at the North Loop entrance to State Street. The State Street facade is emphasized by four columns with lantern capitals and a curved screen backdrop at the roof, and a wall composed of glass and cable-trussed elements. Using natural light and computer-controlled variable lighting modes, the facade will become an architectural embodiment of the energy of Chicago, celebrating the client's product and the building's function. The layering of architectural elements and the interplay of light and colored glass characterize the State Street facade. At ground level, three large display windows will be devoted to permanent exhibits of civic art.

The Lake Street facade is divided by an ornamental stair tower, enclosed in clear glass with stainless steel and painted aluminum frames. At pedestrian level, a series of bronze plaques created by local artists will commemorate events of Chicago's past.

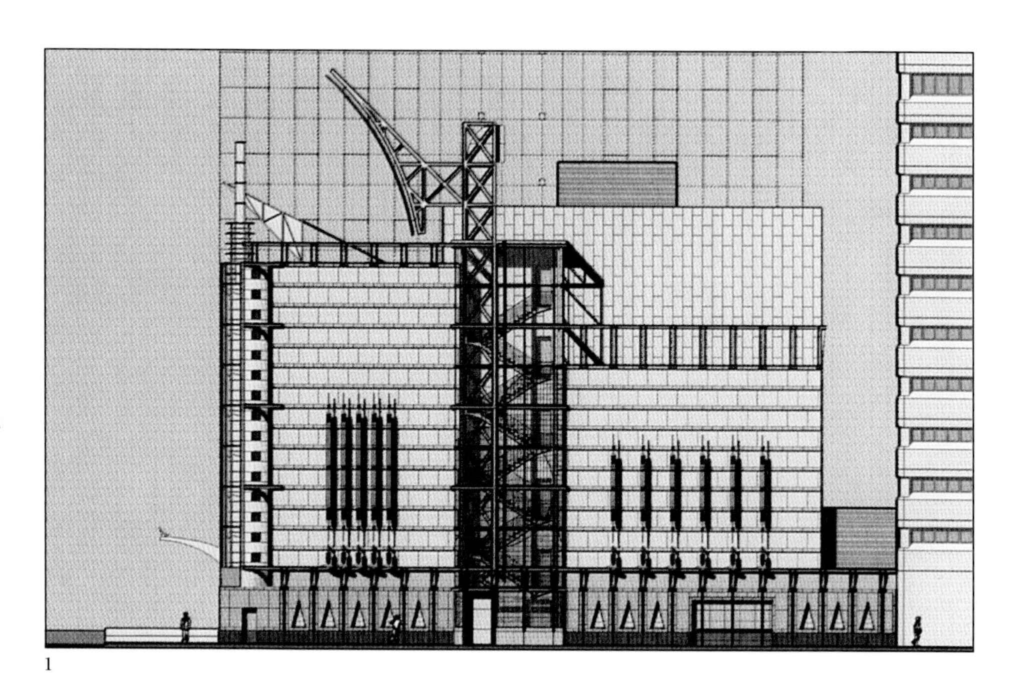

1

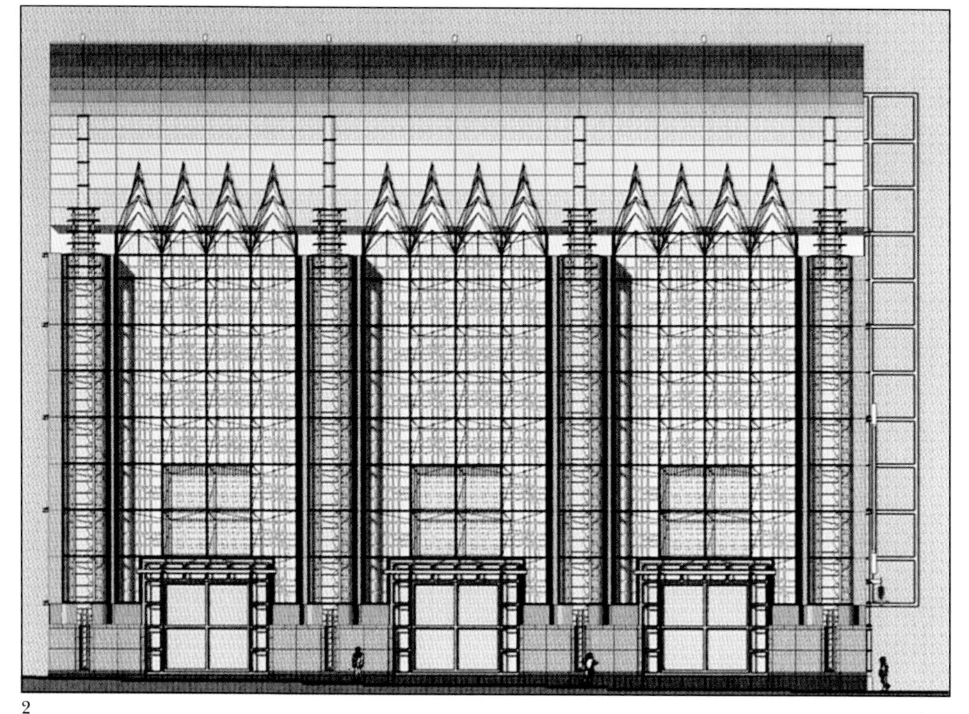

2

1 Lake Street elevation
2 State Street elevation
3–4 State Street concept study

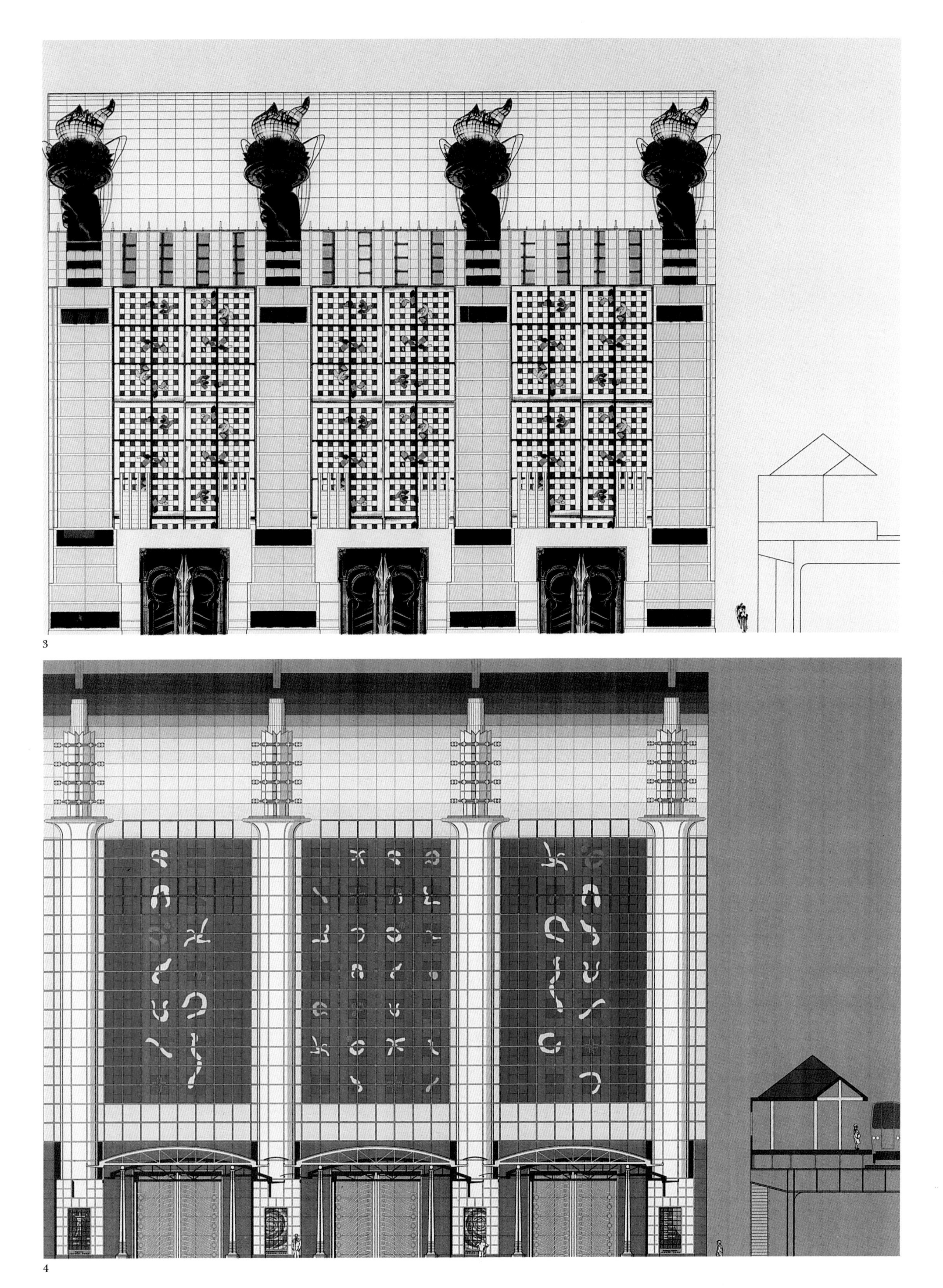

3

4

Selected and Current Works

Design for the Future

Russia Tower

Design/Completion 1992/2001
Moscow, Russia
BDJ International/Fuller
2,600,000 square feet
Structural steel frame with external lateral bracing; concrete core
Glass and aluminum curtain wall

The Russia Tower is envisaged as a symbol of communication between Russia and the rest of the world. As Moscow moves into the 21st century, the Russia Tower will be a significant marker on the skyline. The 126-story tower will be the central element on a city block that is to contain a 400-room hotel with convention facilities, a retail component, and a transportation hall linked to the Moscow Metro. A pedestrian bridge will link the site with the adjacent residential district across the Moscow River.

Together, the various elements of the complex and the bridge create a fortification around the tower, which is sited adjacent to riverfront open space connecting it with the historic center of Moscow and the Kremlin. In contrast to the heaviness of the surrounding elements, the tower's glass and aluminum curtain wall is transparent and light.

1

1 View of tower along Moscow River
2 Aerial view depicting observation level

2

Jin Mao Building

Design/Completion 1993/1998
Shanghai, China
China Shanghai Foreign Trade Centre Co. Ltd
2,852,530 square feet
Steel frame
Glass and metal curtain wall

Located in the Pudong District in Shanghai's Lujiazui Finance and Trade Zone, the project is a multi-use development incorporating offices, a hotel, retail, service amenities, and parking. The tower recalls historic Chinese pagoda forms, with setbacks that create a rhythmic pattern. Its articulated metallic surface captures changes in the light during the day, while at night the shaft and top are illuminated. At 1,380 feet, the tower and its spire are a significant addition to the Shanghai skyline.

The 88-story tower houses hotel and office spaces, with the hotel rooms in the top 38 stories affording impressive views of the city and the surrounding region. Office spaces, in the lower 50 stories, are easily accessed by employees and visitors; parking for 1,200 is located below grade. A seven-story vertical shopping center adjacent to the base of the tower and a landscaped courtyard with reflecting pool and seating at the tower base provide visitors with a peaceful retreat.

1

1 Aerial view
2 Concept view looking east

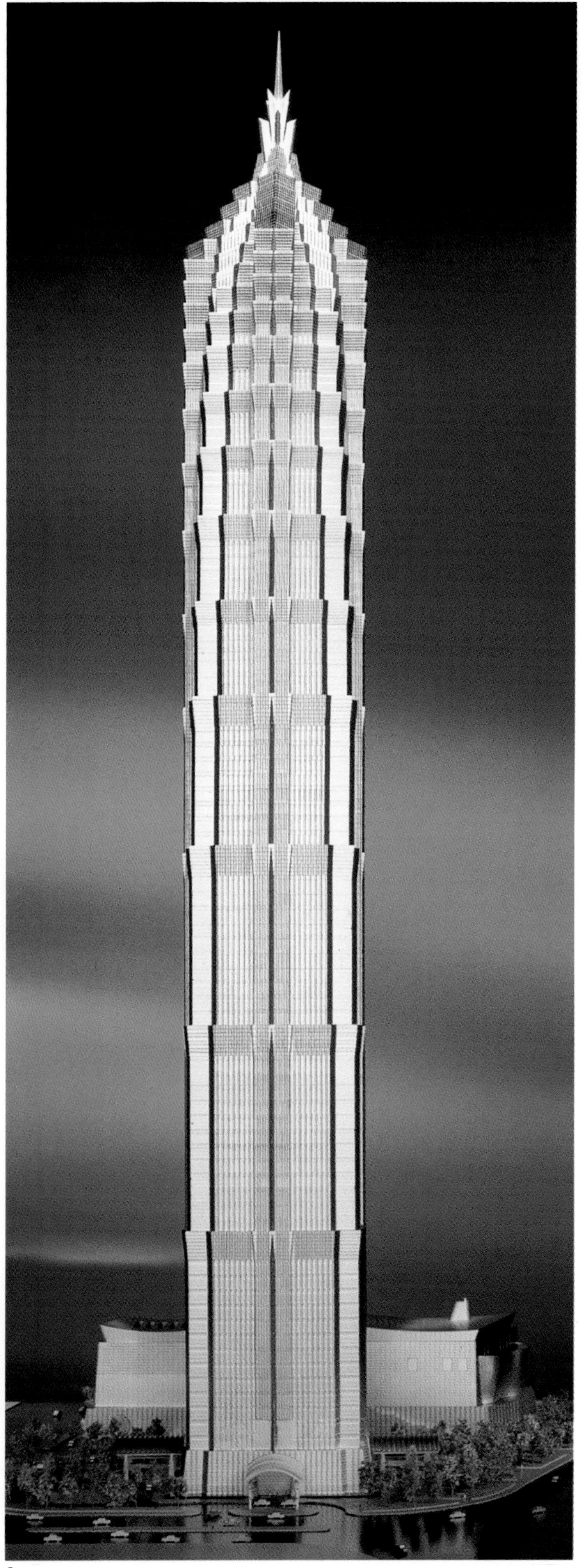

2

International Finance Square

Design/Completion 1993/1996
Guangzhou Tianhe, China
China Entertainment & Land Investment Holdings Ltd
4,500,000 square feet
Steel frame
Granite, glass, metal

Located in the heart of Guangzhou, the tower is the centerpiece of the rapidly developing Tianhe section. The placement of the complex, on an axis with railway and mass transit stations and the Tianhe Sports Center Coliseum, integrates the old city district to the west, the high-technology development zone to the east, and the residential district to the south.

Clad in warm vermilion granite with silver metal mullions, the building soars to over 100 stories, its facade modulated by setbacks that distinguish the base from the shaft. Its octagonal plan is a traditional Chinese symbol of good fortune. A 22-story central atrium brings natural light to the residential units on the top floors of the tower. The extended 12-story base and adjacent buildings include office space, a 900,000 square foot retail mall, and a 500-room hotel.

The sculptural crown shines during the day and is illuminated at night, creating a beacon on Guangzhou's skyline.

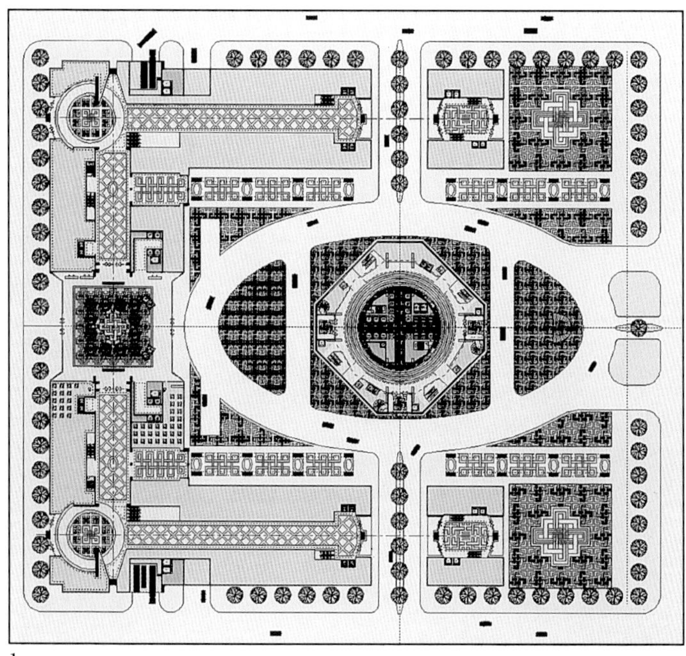

1

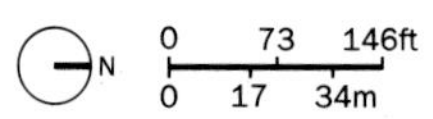

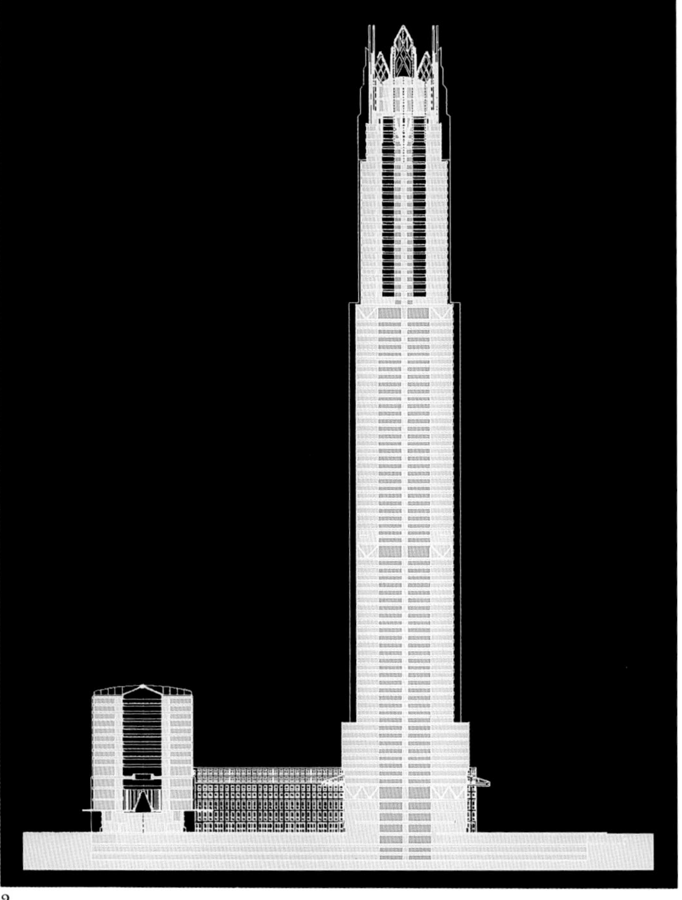

2

1 Ground-floor plan
2 Section through hotel and tower looking west
3 Model: view from north-east

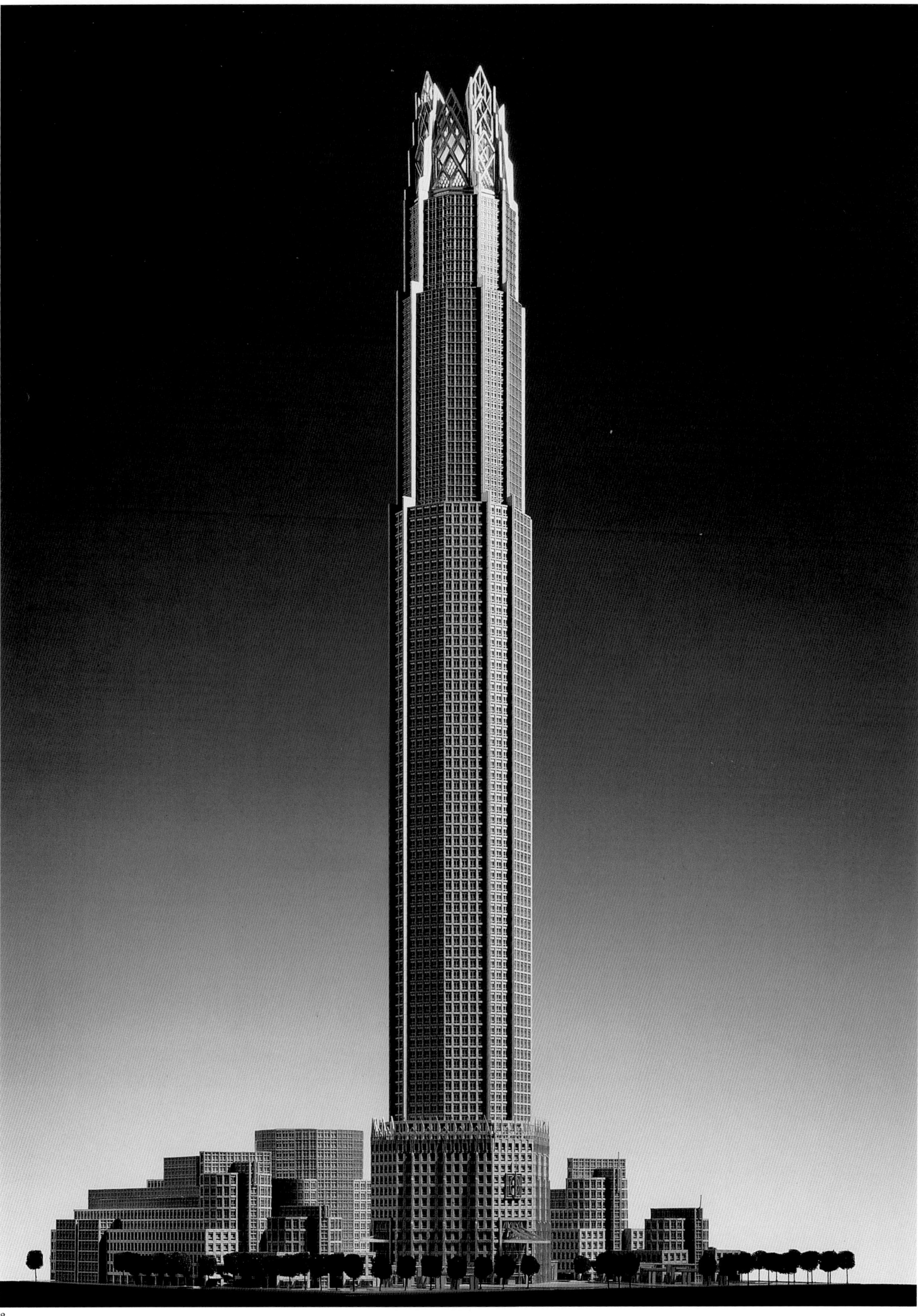

3

World Trade Center Prototype

Design/Completion 1994/1997
Kaneohe, Hawaii, USA
Siegfried Schuster and Arnold Gordon/Mid-Pacific of Hawaii
2,900,000 square feet
Nodally linked structural steel frames
Expressed architectural and structural steel, glass panels

Commercial growth in the 21st century will depend on increased communications and exchange technologies. A series of global merchandise markets capable of housing these developing technologies are essential to the establishment of a functional world trade ethic.

The project is a prototype that accommodates a flexible, multi-level arrangement of expandable trade pavilions for the primary program spaces between two spherical shells. A 500-room hotel, restaurants, and a multi-purpose stadium and performing arts hall are also included in the design.

The use of vision glazing across the interior and exterior shells allows transformative images of reflective architectonic purity by day while permitting full expression of the supporting superstructure by night. Additionally, a central computer will operate both shells as a spherical telescreen monitor capable of mapping world events.

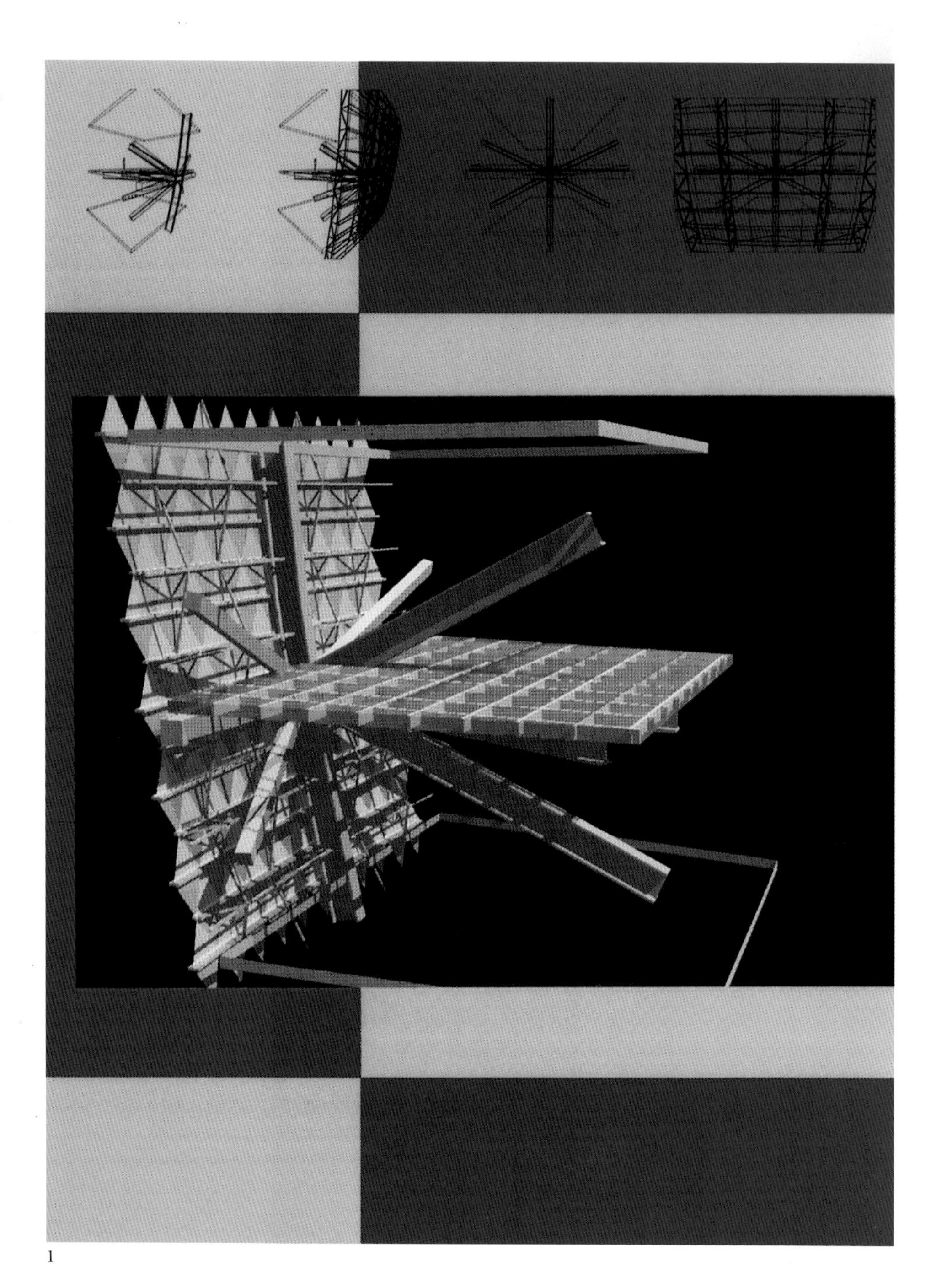

1

1 Detail of curtain wall at typical structural node
2 Cross section through trade floors, hotel and multi-use stadium facilities

Overpage
3 View from the water

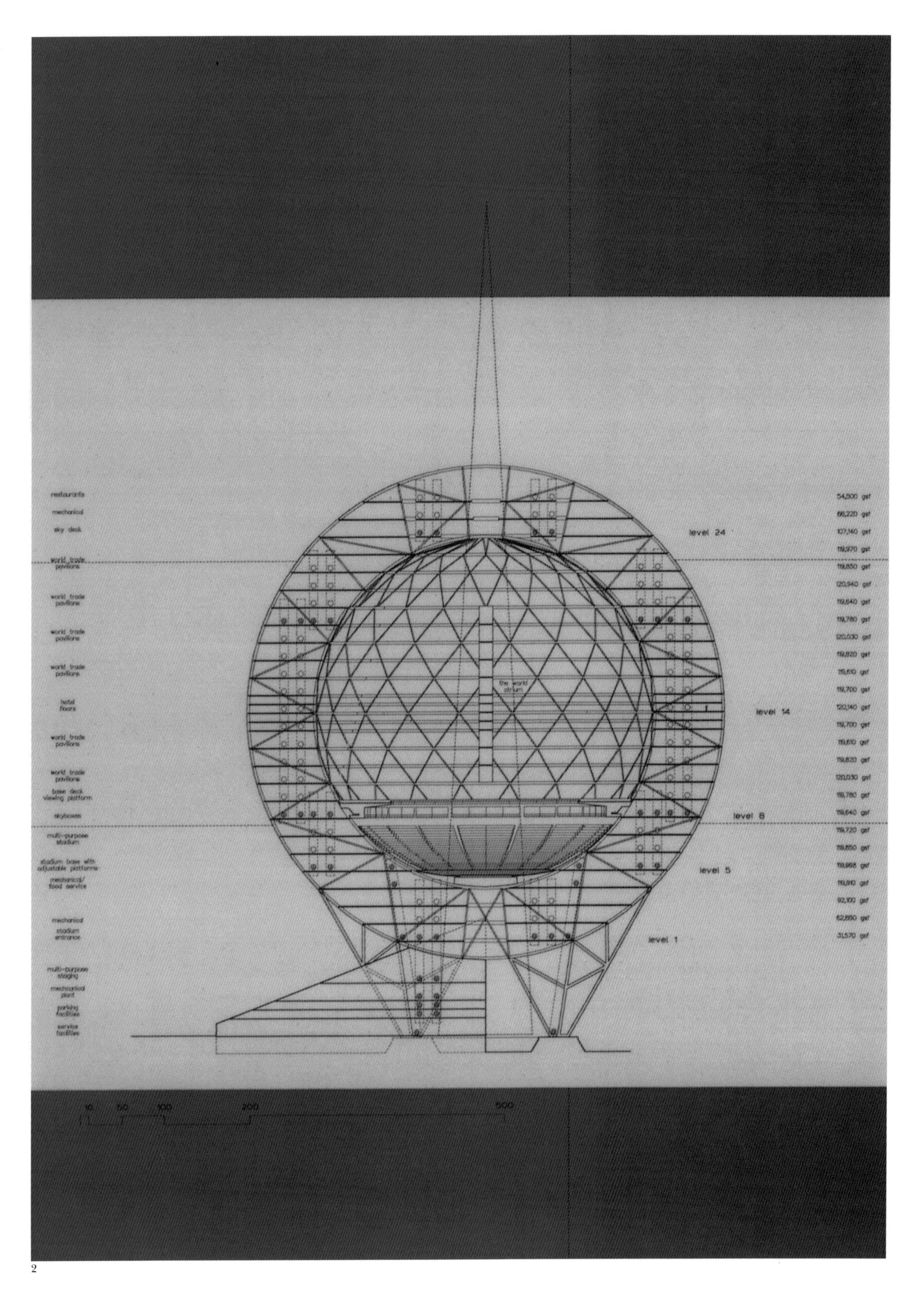

restaurants
mechanical
sky deck
world trade pavilions
world trade pavilions
world trade pavilions
world trade pavilions
hotel floors
world trade pavilions
world trade pavilions
base deck viewing platform
skyboxes
multi-purpose stadium
stadium base with adjustable platforms
mechanical/ food service
mechanical
stadium entrance
multi-purpose staging
mechanical plant
parking facilities
service facilities
the world atrium
level 24
level 14
level 8
level 5
level 1
54,300 gsf
86,220 gsf
107,140 gsf
119,970 gsf
119,850 gsf
120,940 gsf
119,640 gsf
119,780 gsf
120,030 gsf
119,820 gsf
119,610 gsf
119,700 gsf
120,140 gsf
119,700 gsf
119,610 gsf
119,820 gsf
120,030 gsf
119,780 gsf
119,640 gsf
119,720 gsf
119,850 gsf
119,868 gsf
119,910 gsf
92,100 gsf
62,850 gsf
31,570 gsf
10
50
100
200
500

2

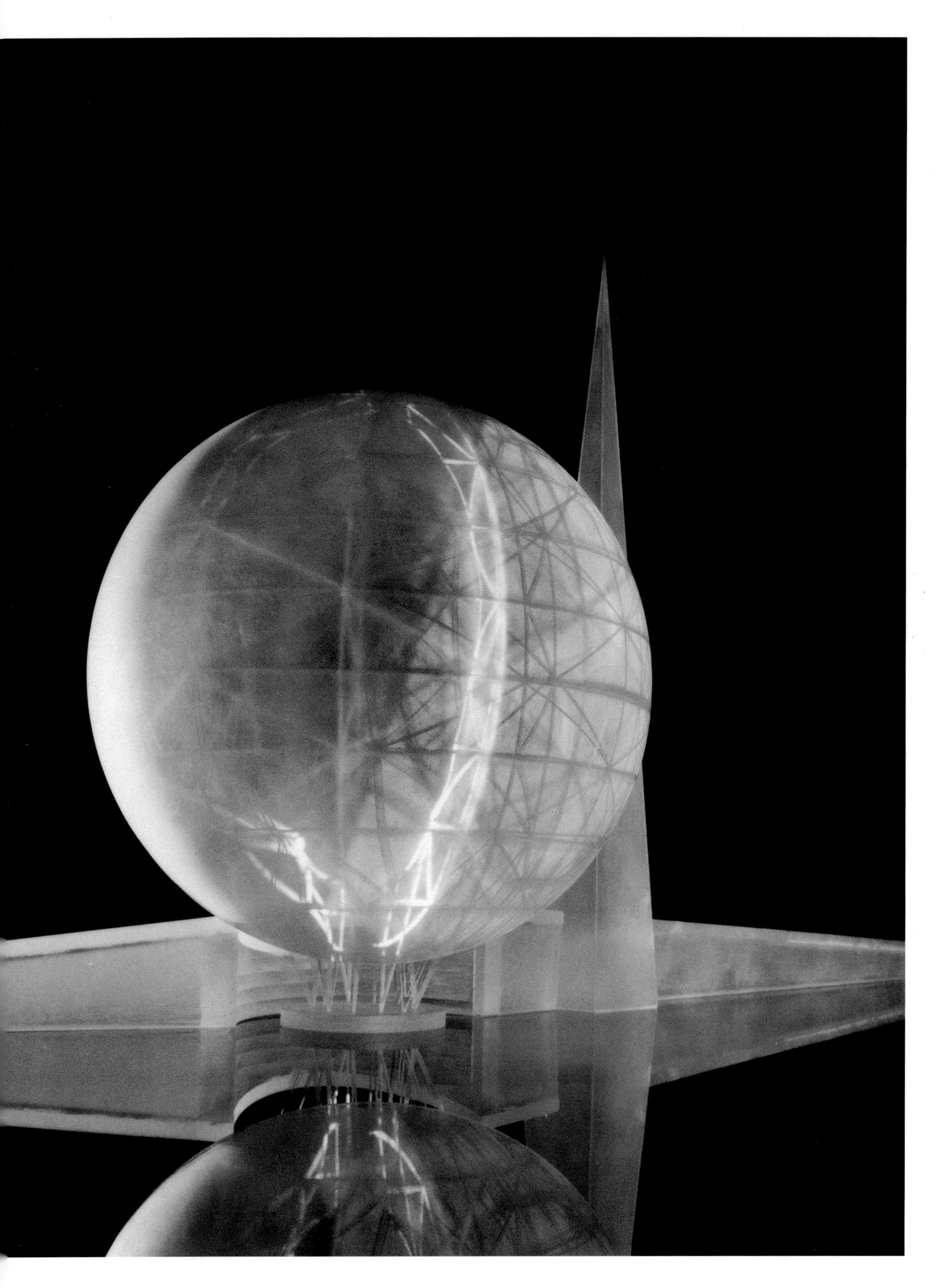

Firm Profile

Partners

Partners 1984–1994

Karen B. Alschuler
Walter W. Arensberg
Robert H. Armsby
Edward C. Bassett
David Magie Childs
James W. Christensen
Richard H. Ciceri
Raymond J. Clark
Walter H. Costa
Michael Damore
Raul de Armas
James R. DeStefano
Robert Diamant
Lawrence S. Doane
William M. Drake
Peter G. Ellis
Thomas J. Eyerman
Richard C. Foster
Thomas K. Fridstein
Richard A. Giegengack
Marc E. Goldstein
Kimbal T. Goluska
Joseph A. Gonzalez
Ted J. Gottesdiener
Bruce J. Graham
Parambir S. Gujral
Robert A. Halvorson
Craig W. Hartman
Alan D. Hinklin
Robert P. Holmes
Peter Hopkinson
Robert A. Hutchins
Srinivasa Iyengar
Marilyn Jordan Taylor
Roger G. Kallman
Richard C. Keating
D. Stanton Korista
John L. Kriken
Diane Legge-Lohan
Richard E. Lenke
Albert Lockett
Peter J. Magill
Warren J. Mathison
Jeffrey J. McCarthy

Partners at 1996 AIA Firm Award Exhibit, Minneapolis, Minnesota

Michael Anthony McCarthy
John O. Merrill, Jr.
Leon Moed
Larry Oltmanns
Maris Peika
David A. Pugh
Gene J. Schnair
Roger M. Seitz
Adrian Devaun Smith
Donald C. Smith
Kenneth A. Soldan
Douglas F. Stoker
Craig W. Taylor
Richard F. Tomlinson, II
Robert Turner
Robert L. Wesley
Gordon L. Wildermuth
John Harbit Winkler
Carolina Y.C. Woo

Awards Worldwide

Since its inception, SOM has been the recipient of over 600 awards. These cover a range of fields including design, planning, interiors, structural design and energy-efficiency, as well as general firm awards. Listed here are some of the firm's more recent awards, chosen from over 500 projects, which refer only to the projects incorporated in this publication.

Firm Awards

1988
For Outstanding Achievement in Design for the Government of the United States of America.

1988
The Tenth Pritzker Architecture Prize, awarded to Gordon Bunshaft.

1985
President of United States Award for Design Excellence, November 10, 1988, President Ronald Reagan.

1983
Gold Medal of Honor
American Institute of Architects, awarded to Nathaniel A. Owings.

1962
Architectural Firm Award
American Institute of Architects.

1960
Landscape Architecture Award (with Isamu Noguchi)
Architectural League.

1957
Gold Medal of Honor
American Institute of Architects, awarded to Louis Skidmore.

Project Awards

New City Architecture Award
Worshipful Company of
Chartered Architects
Ludgate Office Complex
London, England
1994

Honor Award, Distinguished Building Category
Chicago Chapter,
American Institute of Architects
Ludgate, 10 Fleet Place
London, England
1994

Civic Trust Award
Ludgate Office Complex
London, England
1994

National Honor Award, Architecture
American Institute of Architects
Rowes Wharf
Boston, Massachusetts
1994

Unbuilt Design Award, Citation of Merit
Chicago Chapter,
American Institute of Architects
Global Communications Plaza
Rinkutown, Korea
1993

Design Award, Recognition of Health Facilities
New York Chapter,
American Institute of Architects
Columbia-Presbyterian Medical Center
The Milstein Hospital Building
New York, New York
1993

Distinguished Architecture Citation
New York City Chapter,
American Institute of Architects
Transitional Housing for the Homeless
New York, New York
1992

Merit Award for Architecture
Washington, DC Chapter,
American Institute of Architects
100 East Pratt Street
Washington, DC
1992

"Beautification Awards"
For New Commercial High Rise
and Frank Stella mural
Los Angeles Area Chamber of Commerce
Gas Company Tower
Los Angeles, California
1992

Design Award
Texas Society of Architects
Interior Architecture
Solana Marriott Hotel
Westlake, Texas
1992

"Best in Hotel Design" Award
Interiors magazine
Solana Marriott Hotel
Westlake, Texas
1992

Design Award
Interfaith Forum on Religion,
Art and Architecture
Islamic Cultural Center
New York, New York
1992

14th Annual Interiors Award
Interiors magazine
Islamic Cultural Center
New York, New York
1992

Gold Medal Award for Excellence in Masonry Design
Illinois Indiana Masonry Council
AT&T Corporate Center
Chicago, Illinois
1992

British Council Award
British Council for Offices Conference,
Bristol
Broadgate, Exchange House, Phase 11
London, England
1992

First Prize, European Commercial Property Development Awards
Canary Wharf
London, England
1992

Interiors Award
International Design magazine
Solana Marriott Hotel
Westlake, Texas
1991

Honor Award, Interior Design
California Council,
American Institute of Architects
Solana Marriott Hotel
Westlake, Texas
1991

Financial Times Architecture Work Award
Highly Commended
Broadgate, Exchange House, Phase 11
London, England
1991

ECCS European Award for Steel Structures
Broadgate, Exchange House, Phase 11
London, England
1991

Gold Medal, City of London Considerate Contractor Scheme
Broadgate, Phases 5–14
London, England
1991

Art and Work Award
Highly Commended for Howard
Hodgkin's "Wave"
Broadgate Club, Phase 5
London, England
1991

Distinguished Building Award
Certificate of Merit (Unbuilt Design)
Chicago Chapter,
American Institute of Architects
Bayer Milan Office Building
Milan, Italy
1991

First Place, Modern Healthcare Magazine Award Competition
Columbia-Presbyterian Medical Center
The Milstein Hospital Building
New York, New York
1991

Award for Innovation in Building Design and Construction
Progressive Architecture
Broadgate, Exchange House, Phase 11
London, England
1990

National Drywall Award
Broadgate, Phases 5–14
London, England
1990

Dutch National Steel Award
Broadgate, Exchange House, Phase 11
London, England
1990

Construction Achievement Award
Broadgate, Phases 5–14
London, England
1990

Gold Medal, City of London Considerate Contractor Scheme
Broadgate, Phases 5–14
London, England
1990

British Construction Industry Awards
Highly Commended (Building Category)
Broadgate, Exchange House, Phase 11
London, England
1990

Bayer Award of Excellence for Colour in Architecture
Broadgate, Broadwalk House, Phase 9/10
London, England
1990

Distinguished Building Award
Chicago Chapter,
American Institute of Architects
Rowes Wharf
Boston, Massachusetts
1990

Award of Excellence
National Commercial Builder's Council
Rowes Wharf
Boston, Massachusetts
1990

Design Award
Precast/Prestressed Concrete Institute
Aurora Municipal Justice Center
Aurora, Colorado
1990

Best of Competition
Contract Design Competition
Institute of Business Designers/
Interior Design magazine
Latham & Watkins
Los Angeles, California
1990

Beautification Award
(New Commercial Low Rise)
Los Angeles Area Chamber of Commerce
Columbia Savings & Loan Office Building
Beverly Hills, California
1990

Contract Design Competition Top 10
Interior Design magazine
Sun Bank, NA
Orlando, Florida
1989

ULI Award for Excellence
(Large-Scale Urban Development/
Mixed Use)
Urban Land Institute
Rowes Wharf
Boston, Massachusetts
1989

PCI Professional Design Award
Prestressed Concrete Institute
Rowes Wharf
Boston, Massachusetts
1989

Grand Award
Builder magazine
Rowes Wharf
Boston, Massachusetts
1989

National Citation for Excellence in Urban Design
American Institute of Architects
Rowes Wharf
Boston, Massachusetts
1989

Honor Award
California Council,
American Institute of Architects
Pacific Bell Administration Complex
San Ramon, California
1989

Distinguished Detail Award
Honor Award
Chicago Chapter,
American Institute of Architects
United Gulf Bank
Manama, Bahrain
1989

Special Award
Concrete Federation of Holland
Broadgate, Broadwalk House, Phase 9/10
London, England
1989

Merit Award, City of London Considerate Contractor Scheme
Broadgate, Phases 5–14
London, England
1989

First Prize, Building Award Competition
Queens Chamber of Commerce
Citicorp at Court Square
Long Island City, New York
1989

Interiors Award
Interiors magazine
Palio Restaurant
New York, New York
1988

Award for Excellence in Design
Art Commission of the City of New York
Transitional Housing for the Homeless
New York, New York
1988

Urban Design Citation
Boston Chapter,
American Institute of Architects
Rowes Wharf
Boston, Massachusetts
1988

Honor Award for Excellence on the Waterfront
The Waterfront Center
Rowes Wharf
Boston, Massachusetts
1988

National Honor Award
American Institute of Architects
United Gulf Bank
Manama, Bahrain
1988

Outstanding Achievement Award
Interior Design magazine
United Gulf Bank
Manama, Bahrain
1988

Interior Architecture Award
Chicago Chapter,
American Institute of Architects
United Gulf Bank
Manama, Bahrain
1988

Distinguished Building Award
Chicago Chapter,
American Institute of Architects
United Gulf Bank
Manama, Bahrain
1988

Citation of Merit
Chicago Chapter,
American Institute of Architects
United Gulf Bank
Manama, Bahrain
1988

First Place Award
Restaurant Hospitality magazine
Palio Restaurant
New York, New York
1987

Honor Award
San Francisco Chapter,
American Institute of Architects
388 Market Street
San Francisco, California
1987

Energy Conservation Award
First Place (New Commissioned Buildings)
ASHRAE
Citicorp at Court Square
Long Island City, New York
1987

Distinguished Building Award
Chicago Chapter,
American Institute of Architects
Art Institute of Chicago, Renovation of
Second Floor Galleries
Chicago, Illinois
1987

Architectural Award of Excellence
American Institute of Steel Construction
McCormick Place Expansion
Chicago, Illinois
1987

Architectural Award of Excellence
American Institute of Steel Construction
Michigan Power, Morrow Dam
Kalamazoo County, Michigan
1987

ASHRAE Energy Award
ASHRAE
303 West Madison
Chicago, Illinois
1986

Urban Design & Planning Citation
Progressive Architecture
Northwest Frontier Province Agricultural
University Master Plan
Peshawar, Pakistan
1985

First Place Winner,
ASHRAE Energy Awards
Illinois Chapter, ASHRAE
Asian Development Bank
Manila, Philippines
1985

Architectural Design Citation
Progressive Architecture
303 West Madison
Chicago, Illinois
1985

Design Excellence Award
Washington, DC Chapter,
American Institute of Architects
Baltimore Pennsylvania Station
Baltimore, Maryland
1985

Honorable Mention
(Adaptive Re-Use Category)
Interior Design Competition
Institute of Business Designers/
Interior Design magazine
Baltimore Pennsylvania Station
Baltimore, Maryland
1984

First Award for Achievement of Excellence
in Historic Preservation and Architecture
Washington, DC Chapter,
American Institute of Architects
Baltimore Pennsylvania Station
Baltimore, Maryland
1984

Federal Design Achievement Award
Baltimore Pennsylvania Station
Baltimore, Maryland
1984

First Place, Energy Conservation Awards
(New Multi-Use Building)
Rowes Wharf
Boston, Massachusetts
1984

Architectural Design Citation
Progressive Architecture
United Gulf Bank
Manama, Bahrain
1984

Architectural Design Citation
Progressive Architecture
Providence Station
Providence, Rhode Island
1983

Architectural Design Citation
Progressive Architecture
388 Market Street
San Francisco, California
1983

Merit Award
Portland Chapter,
American Institute of Architects
Islamic Cultural Center
New York, New York
1982

Best Structure
Structural Engineers Association of Illinois
Hubert H. Humphrey Metrodome
Minneapolis, Minnesota
1982

Merit Award for Achievement of Excellence in Historic Preservation and Architectural Design
Washington, DC Chapter,
American Institute of Architects
Baltimore Pennsylvania Station
Baltimore, Maryland
1981

Merit Award
American Society of Landscape Architects
Art Institute of Chicago Centennial Addition
Chicago, Illinois
1978

Honor Award
American Institute of Architects
Art Institute of Chicago Centennial Addition
Chicago, Illinois
1978

Bibliography

Books (1982–1994)

Adler, Jerry. *High Rise.* New York: Harper Collins, 1993, pp. 69, 71–75, 84–98, 200, 122–124, 140, 145–151, 155–156, 158, 160, 172–82, 196–198, 210, 228, 230, 235, 244–245, 247–249, 251, 274–276, 280, 285, 291, 315, 321, 325, 329, 356.

Architecture Chicago, Vol. 7: Alternative Visions. Chicago: Chicago Chapter AIA, 1989.

Architecture Contemporaine. Lausanne, Switzerland: Editions Anthony Krafft, 1983, p. 91 (King Abdul Aziz International Airport).

Arnell, Peter & Bickford, Ted (eds). *Southwest Center: The Houston Competition.* New York: Rizzoli, 1983.

Attoe, Wayne & Logan, Donn. *American Urban Architecture: Catalysts in the Design of Cities.* Berkeley: University of California Press, 1989, pp. 49, 82, 94, 95, 139, 151.

Austin, Richard L. *Adaptive Reuse: Issues and Case Studies in Building Preservation.* New York: Van Nostrand Reinhold, 1988 (Wacker Plaza).

Bach, Ira S. & Wolfson, Susan. *Chicago on Foot: Walking Tours of Chicago's Architecture.* 5th edn. Chicago: Chicago Review Press, 1994.

Bednar, Michael J. *Interior Pedestrian Places.* New York: Whitney, 1989, pp. 40, 119, 155, 198, 208.

Billington, David P. & Goldsmith, Myron (eds). "Technique and Aesthetics in the Design of Tall Buildings." *Proceedings of the Fazlur R. Khan Session on Structural Expression in Buildings.* Annual Fall Meeting, Houston, Texas: American Society of Civil Engineers, 1983.

Blaser, Werner. *Buildings And Concepts.* Basel, Switzerland: Birkhauser, 1986. (Myron Goldsmith).

Bruegmann, Robert (ed.). *Modernism at Mid-century: The Architecture of the United States Air Force Academy.* Chicago: University of Chicago Press, 1994.

Bush-Brown, Albert. *Skidmore, Owings & Merrill: Architecture and Urbanism, 1973–1983.* London: Thames and Hudson, 1984.

Chicago Museum of Science and Industry. *A Guide to 150 Years of Chicago Architecture.* Chicago: Chicago Review Press, 1988, pp. 106–115 (Walter Netsch).

Christ, Ronald & Dollens, Dennis. *New York: Nomadic Design.* Barcelona: Editoriale Gustavo Gili, 1993, pp. 88–89 (Flushing Meadows).

Council on Tall Buildings and Urban Habitat. *Developments in Tall Buildings 1983.* Stroudsburg, Pennsylvania: Hutchinson Ross Publishing Co., 1983.

Curran, Raymond J. *Architecture and the Urban Experience.* New York: Van Nostrand Reinhold, 1983, p. 199 (Marine Midland, Chase Manhattan Bank Plaza).

Current Biography Yearbook. New York: Gale, 1989 (Gordon Bunshaft).

Curtis, William J.R. *Modern Architecture Since 1900.* 2nd edn. London: Phaidon, 1987, pp. 10, 19, 266, 267, 349, 365, 392, 393, 395.

Darragh, Joan (ed.). *The New Brooklyn Museum: The Master Plan Competition.* New York: Rizzoli/Brooklyn Museum, 1988, pp. 122–141.

de Witt, Dennis J. *Modern Architecture in Europe: A Guide to Buildings since the Industrial Revolution.* New York: Dutton, 1987, pp. B8, GB579.

Delehanty, Randolph. *Ultimate Guide San Francisco.* San Francisco: Chronicle Books, 1989, pp. 50, 60, 64, 66–67, 69, 71–72, 74–75, 102.

Diamonstein, Barbaralee. *American Architecture Now II.* New York: Rizzoli, 1985, pp. 120–127.

Diamonstein, Barbaralee. *Landmarks: Eighteen Wonders of the New York World.* New York: Abrams, 1992, p. 142 (Lever House).

Dober, Richard P. *Campus Design.* New York: Wiley, 1992, pp. 92–93, 116, 232, 235.

Drexel, John (ed.). *The Facts on File Encyclopedia of the 20th Century.* New York: Facts on File, 1991, pp. 146–147 (Gordon Bunshaft), p. 833 (SOM).

Dudley, George A. *A Workshop for Peace: Designing the United Nations Headquarters.* New York: Architectural History Foundation/Cambridge, Massachusetts: MIT Press, 1994, p. 353 (Louis Skidmore's role in design of UN; personal profile).

Dunlap, David W. *On Broadway: A Journey Uptown Over Time.* New York: Rizzoli, 1990, pp. 26, 27, 167, 171, 207, 208, 234, 296, 314.

Encyclopaedia Britannica. 15th edn, vol. 13, 1992, pp. 990–991.

Evette, Therese. *L'architecture Tertiaire en Europe et aux Etats-Unis.* Paris: Ministére de l'Equipement, des Transports et du Logement/Primavera Quotidienne, 1992, p. 115 (Irving Trust), p. 117 (Shearson American Express), p. 131 (San Benigno Torre).

Gebhard, David & Winter, Robert. *Los Angeles: An Architectural Guide.* Layton, Utah: Gibbs Smith, 1994.

Goldberger, Paul. *On the Rise: Architecture and Design in a Postmodern Age.* New York: Times Books, 1983, pp. 21, 35, 59, 51, 60, 79, 80, 253, 255.

Graham, Bruce. *Bruce Graham Of SOM.* New York: Rizzoli, 1989.

Gratz, Roberta Brandes. *The Living City.* New York: Simon & Schuster, 1989, p. 297.

Gretes, Frances C. *Gordon Bunshaft: The Life and Work of a Modern Master.* Monticello, Illinois: Vance Bibliographies, 1988.

Gretes, Frances C. *Lever House: An Architectural Landmark: Bibliography Of Sources, 1950–1983.* Monticello, Illinois: Vance Bibliographies, 1988.

Gretes, Frances C. *Skidmore, Owings & Merrill, 1936–1983.* Monticello, Illinois: Vance Bibliographies, 1984.

Hartt, Frederick. *A History of Painting, Sculpture and Architecture.* 3rd edn. New York: Abrams, 1989, p. 975 (Lever House).

Heyer, Paul. *American Architecture: Ideas and Ideologies in the Late Twentieth Century.* New York: Van Nostrand Reinhold, 1993, pp. 53–55, 67, 69–70, 76–78, 80, 95, 97, 198, 215, 217.

Heyer, Paul. *Architects on Architecture: New Directions in America.* New York: Van Nostrand Reinhold, 1993, pp. 40, 92, 113, 290–291, 322, 352, 362–377.

Huxtable, Ada Louise. *Architecture Anyone? Cautionary Tales of the Building Art.* New York: Random House, 1986, pp. 294–296.

Interior Designers of the United States. Melbourne, Australia: Images Publishing Group, 1991, pp. 80–82.

Jencks, Charles. *The New Moderns.* New York: Rizzoli, 1990, p. 81 (Wills Tobacco Factory), p. 82 (Hancock Center), p. 91 (LBJ Library), p. 91 (Hirshhorn Museum), p. 170 (W.R. Grace).

Jodido, Philip. *Contemporary American Architects.* Köln: Benedikt Taschen, 1993, p. 27 (Gordon Bunshaft).

Jones, Edward & Woodward, Christopher. *A Guide to the Architecture of London.* 2nd edn. London: Thames & Hudson, 1992, pp. K165, K170, R30, T33.

Karaman, Aykut. *The Ecology of the Built Environment: Toward Designing for 'Genius Loci'.* Philadelphia: University of Pennsylvania, 1983.

Katz, Peter. *The New Urbanism: Toward an Architecture of Community.* New York: McGraw-Hill, 1992, pp. 164-167 (Atlantic Center).

Kennedy, Roger G. (ed.). *American by Design.* New York: Oxford University Press, 1987, pp. 123, 127, 187, 245.

Kennedy, Roger G. (ed.). *Smithsonian Guide to Historic America: The Mid-atlantic States.* New York: Stewart Tabori & Chang, 1989, pp. 87–88 (Lever House).

Kidder Smith, G.E. *Looking at Architecture.* New York: Abrams, 1990, p. 166 (Chase).

Kostof, Spiro. *The City Assembled.* Boston: Little, Brown & Co., 1992, pl. 5 (Rowes Wharf).

Krinsky, Carol Herselle. *Gordon Bunshaft of Skidmore, Owings & Merrill.* New York: Architectural History Foundation/ Cambridge, Massachusetts: MIT Press, 1988.

Lampugnani, Vittorio Magnago. *Encyclopedia of 20th Century Architecture.* New York: Abrams, 1984, pp. 43, 44, 83, 125, 163, 164, 181, 210, 213, 305–308, 310, 344, 350–352.

Larson, George A. & Pridmore, Jay. *Chicago Architecture and Design.* New York: Abrams, 1993.

Laventhol & Horwath. *Convention Centers, Stadiums & Arenas.* Washington, DC: Urban Land Institute, 1989, pp. 98–100 (Humphry Metrodome).

Lee, Marshall (ed.). *Art at Work: The Chase Manhattan Collection.* New York: Dutton, 1984.

Letts, Vanessa. *Cadogan City Guides: New York.* Old Saybrook, Connecticut: Globe Pequot Press, 1993, pp. 86–87, 106–107.

MacDonald, Angus J. *Structure and Architecture.* Oxford: Butterworth, 1994, pp. 93–94 (Hancock Center, Sears Tower).

Malik, Arshad Salim. *The Effects of International Governmental Donor Agencies on Institution-building: A Case Study of United States Agency for International Development's Assistance to the North-west Frontier Province Agricultural University, Peshawar (Pakistan).* Urbana-Champaign, Illinois: University of Illinois, 1993.

Master Builders: A Guide to Famous American Architects. Washington, DC: National Trust for Historic Preservation, 1985, pp. 9, 10, 160–163.

McGrew, Patrick. *Landmarks of San Francisco.* New York: Abrams, 1991, pp. 10, 260–261 (Crown Zellerbach).

Moffett, Noel. *The Best of British Architecture, 1980–2000.* London: E & F.N. Spon, 1993, p. 59 (Canary Wharf).

Morgan, Ann L. (ed.). *Contemporary Architects.* 2nd edn. New York: St James Press, 1987, pp. 134–136 (Gordon Bunshaft), pp. 332–334 (Myron Goldsmith), p. 342 (Bruce Graham), pp. 650–651 (Walter Netsch), pp. 846–848 (SOM).

Morrone, Francis. *Architectural Guidebook to New York City.* Layton, Utah: Gibbs Smith, 1994, pp. 26–29, 58, 61, 116, 145, 167, 199, 203, 211, 245, 253, 254, 258, 263, 265, 270, 318.

Murray, Peter. *SOM: Reality Before Reality.* Milan: Edizioni Tecno, 1990.

Phillips, Alan. *The Best in Leisure and Public Architecture.* Mies, Switzerland: Rotovision, 1993 (Broadgate).

Placzek, Adolf K. (ed.). *Macmillan Encyclopedia of Architects.* New York: Macmillan, 1982, vol. 1, pp. 332–334 (Gordon Bunshaft), vol. 3, pp. 77– 80 (SOM).

Plotch, Batia (ed.). *New York Walks.* New York: Holt, 1992, pp. 167, 180, 198–199.

Plunz, Richard. *A History of Housing in New York.* New York: Columbia University Press, 1990, pp. 264–266, 282, 287–288.

Raeburn, Michael (ed.). *Architecture of the Western World.* London: Popular Press, 1988, pp. 247, 256, 275.

Rajs, Jake. *Manhattan: An Island in Focus.* New York: Rizzoli, 1985, p. 52 (140 Broadway).

Reynolds, Donald Martin. *The Architecture of New York City: Histories and Views of Important Structures, Sites and Symbols.* Rev. edn. New York: Wiley, 1994, pp. 171, 180.

Robinson, Cervin. *Architecture Transformed: A History of the Photography of Buildings from 1835 to the Present.* New York: Architectural History Foundation/Cambridge, Massachusetts: MIT Press, 1987, pp. 146, 149, 156.

Roth, Leland M. *Understanding Architecture.* New York: Harper Collins, 1993, pp. 19, 20, 67, 131, 472.

Rowe, Peter G. *Making a Middle Landscape.* Cambridge: MIT Press, 1991, pp. 128, 153, 154, 158, 159, 160.

Rowe, Peter G. *Modernization and Housing.* Cambridge, Massachusetts: MIT Press, 1993, p. 192 (North Harlem Public Housing).

Russell, Beverly. *Architecture & Design, 1970–1990: New Ideas in America.* New York: Abrams, 1989, pp. 37 (Lever House, Manufacturers Hanover Trust), p. 56 (Merrill Lynch Trading Room), pp. 118–119 (Columbus Circle).

Sabbagh, Karl. *Skyscraper.* New York: Viking, 1990.

Saliga, Pauline A. *The Sky's the Limit: A Century of Chicago Skyscrapers.* New York: Rizzoli, 1990.

Sanford, John Douglas. *The Gallery Architects: Edward B. Green and Gordon Bunshaft.* Buffalo, New York: Albright-Knox Art Gallery, 1987.

Saxon, Richard. *Atrium Buildings: Development and Design.* New York: Van Nostrand Reinhold, 1983, pp. 50, 114, 129, 130.

Scully, Vincent. *Architecture: The Natural and the Manmade.* New York: St Martin's Press, 1991, pp. 2, 3, 205–206 (Lever House).

Sharp, Dennis (ed.). *The Illustrated Encyclopedia of Architects and Architecture.* New York: Whitney/Quarto, 1991, pp. 48, 80, 106, 129, 142–143, 244.

Shay, James. *New Architecture San Francisco.* San Francisco: Chronicle Books, 1989, pp. 114–117.

Sinkevitch, Alice. *AIA Guide to Chicago.* New York: Harcourt Brace & Co., 1993.

Slavin, Maeve. *Davis Allen: Forty Years of Interior Design at Skidmore, Owings & Merrill.* New York: Rizzoli, 1990.

Southard, Edna, C. *Living with Art, Two: The Collection of Walter and Dawn Clark Netsch.* Oxford, Ohio: Miami University Art Museum, 1983.

Steele, James. *The Contemporary Condition: Los Angeles Architecture.* London: Phaidon Press, 1993, p. 302 (Crocker Center).

Stern, Robert A.M. *Modern Classicism.* New York: Rizzoli, 1988, pp. 174, 280–281.

Stern, Robert A.M. *Pride of Place.* Boston: Houghton Mifflin, 1986, pp. 4, 38, 72–73, 283, 289, 292–293.

Stoller, Ezra. *Modern Architecture.* New York: Abrams, 1990, pp. 7, 69–72, 83, 85–114.

Toy, Maggie (ed.). *World Cities Los Angeles.* London: Academy Editions, 1994, p. 162 (Gas Company Tower), p. 166 (Wells Fargo).

Trachtenberg, Marvin. *Architecture from Prehistory to Post-modernism: The Western Tradition.* New York: Abrams, 1986, pp. 545–546 (Lever House, Hancock Center, Chicago).

Trench, Richard. *Philip's London: Architecture, History, Art.* London: George Philip, 1991, p. 11 (Canary Wharf).

Turner, Paul Venable. *Campus: An American Planning Tradition.* New York: Architectural History Foundation/ Cambridge, Massachusetts: MIT Press, 1984, pp. 264, 266, 268, 269, 271, 274, 278, 279, 280.

Weinreb, Matthew. *London Architecture: Features and Facades.* London: Phaidon, 1993, pp. 2, 24–25, 60–61 (Canary Wharf).

White, Norval. *New York: A Physical History of New York.* New York: Atheneum, 1987, pp. 137, 160, 180, 182, 185, 192, 198, 200–202, 212, 224.

Whyte, William H. *City: Rediscovering the Center.* New York: Doubleday, 1988, pp. 269, 285.

Willensky, Elliot & White, Norval. *AIA Guide to New York.* 3rd edn. Washington, DC: American Institute of Architects, 1988.

Williamson, Roxanne Kuter. *American Architecture and the Mechanics of Fame.* Austin: University of Texas Press, 1991, pp. 3, 24, 27, 66, 96, 113, 114, 119, 125, 127, 128, 136, 188, 204, 205, 249.

Wolfe, Gerard R. *New York: A Guide to the Metropolis.* 2nd edn. New York: McGraw-Hill, 1994, pp. 12–13, 56, 67, 306, 310, 315, 347–348.

Woodbridge, Sally B. *San Francisco Architecture: The Illustrated Guide to over 1,000 of the Best Buildings, Parks and Public Artworks in the Bay Area.* San Francisco: Chronicle Books, 1992.

Wright, Carol von Pressentin. *Blue Guide New York.* 2nd edn. New York: Norton, 1991, pp. 115, 132, 135, 138, 156, 293, 304, 316, 326, 333, 338, 381, 385, 390, 392, 432, 497, 504, 571, 585, 639, 684, 704, 705.

Yee, Roger & Gustafson, Karen. *Corporate Design.* New York: Whitney, 1983 (Weyerhaeuser).

Articles

"A Railroad Station for Providence, RI." *Progressive Architecture* (vol. 64, no. 1, January 1983): pp. 100–101.

Al-Towim, Yousef et al. "Jeddah Dossier." *Albenaa* (special issue, no. 25, October/ November 1985): pp. 22–57 (National Commercial Bank, Jeddah).

Arcidi, Philip "10 Ludgate." *Progressive Architecture* (vol. 74, no. 8, August 1993): p. 88.

Arcidi, Philip. "Hospitals Made Simple." *Progressive Architecture* (vol. 73, no. 3, March 1992): pp. 86–95 (Milstein Hospital Building).

Arnaboldi, Mario Antonio et al. "La Teoria delle 'Catastrofi'." (The Barcelona '92 Olympics.) *l'Arca* (no. 59, April 1992): pp. 48–59, 68–81, 86–89, 100–103 (Hotel de las Artes).

Arnaboldi, Mario Antonio et al. "Le Torri." (Towers.) *l'Arca* (special issue, no. 72, June 1993): pp. 2–85, 97.

Aymonino, Carlo et al. "Who Designs the City?" *Zodiac* (special issue, no. 5, 1991): pp. 4–211 (Canary Wharf).

"Award for Planning Achievement 1992: London's Broadgate Development Wins the Silver Jubilee Cup." *Planner* (vol. 78, no. 22, December 11, 1992): pp. 15–17.

Bachmann, Wolfgang et al. "Chicago." *Bauwelt* (special issue, vol. 80, no. 40/41, October 27, 1989): pp. 1920–1958 (Citifront Center, AT&T, USG Building).

Barreneche, Raul. "Dulles Airport Expansion, Washington, DC." *Architecture* (AIA, vol. 82, no. 11, November 1993): pp. 40–41.

Betsky, Aaron. "Fire and Ice." *Architectural Record* (vol. 180, no. 5, May 1992): pp. 102–111 (Gas Company Tower).

Binney, Marcus. "Canary Wharf, etc." *Architectural Design* (vol. 59, no. 5/6, May/ June 1989): pp. iv–viii.

Bradshaw, Martin. "Civic Trust Awards." *Building* (special issue, vol. 259, no. 7843 (19) supplement, May 13, 1994): pp. 4–50.

Branch, Mark Alden. "Religious Buildings." *Progressive Architecture* (vol. 71, no. 12, December 1990): pp. 78–85 (Islamic Cultural Center, New York).

Byrne, Timothy. "Home of the High Rise: the Architecture of Skidmore Owings & Merrill." *Crit* (no. 13, Fall 1983): pp. 42–46.

Camacho, Carlos Cubillos. "Skidmore Owings & Merrill: Recent Work." *Proa* (special issue, no. 329, April 1984): pp. 18–59.

Campbell, Robert et al. "Building Types Study: 698. Social Housing." *Architectural Record* (vol. 180, no. 7, July 1992): pp. 69–117 (Transitional Housing, New York).

Campbell, Robert et al. "Structure." *Architecture* (AIA, special issue, vol. 77, no. 3, March 1988): pp. 67–150 (John Hancock Tower, Boston; McCormick Place, Chicago; 388 Market Street, San Francisco).

Campbell, Robert et al. "The Eleventh Annual Review of New American Architecture." *Architecture* (AIA, special issue, vol. 77, no. 5, May 1988): pp. 58–194.

Campbell, Robert et al. "The Sixth Annual Review of New American Architecture." *AIA Journal* (special issue, vol. 72, no. 5, May 1983): pp. 149–359.

Cannon-Brookes, Peter. "Renovation and Reinstallations at the Art Institute of Chicago." *International Journal of Museum Management & Curatorship* (vol. 6, no. 3, September, 1987): pp. 307–312.

"Checkpoint Charlie Project Embraces Berlin Wall." *Architectural Record* (vol. 180, no. 12, December 1992): p. 11.

Chevin, Denise. "Slimming Lessons." *Building* (vol. 257, no. 7730 (5), January 31, 1992): pp. 40–45 (Ludgate).

Coomber, Matthew et al. "Technology Special: Concrete." *Building* (vol. 255, no. 7642 (14), April 6, 1990): pp. 51–74 (Broadgate).

Cornuejols, Michel et al. "Arquitectura Metalica." (Metal architecture.) *Proa* (special issue, no. 400, April 1991): pp. 10–57 (Broadgate).

Davidson, Cynthia Chapin. "Urbane Renewal." *Inland Architect* (vol. 33, no. 5, September/October 1989): pp. 43–49 (Rowes Wharf).

Davidson-Powers, Cynthia et al. "SOM: 50 and Counting." *Inland Architect* (vol. 31, no. 2, March/April, 1987): pp. 28–53.

Davidson-Powers, Cynthia. "An Unconventional Tribute." *Inland Architect* (vol. 32, no. 5, September/October 1988): pp. 42–46 (McCormick Place Exposition Center).

Davidson-Powers, Cynthia. "Seeking New Identities: Two Projects by Skidmore Owings & Merrill." *Inland Architect* (vol. 32, no. 3, May/June 1988): pp. 54–59 (225 West Washington, 303 West Madison, Chicago).

Dean, Andrea Oppenheimer et al. "Adding New to Old." *Architecture* (AIA, special issue, vol. 79, no. 11, November 1990): pp. 65–101 (Solana Marriott Hotel, Texas).

Dean, Andrea Oppenheimer et al. "American Architects Abroad." *Architecture* (AIA, special issue, vol. 79, no. 9, September 1990): pp. 57–97 (Canary Wharf, Broadgate).

Dean, Andrea Oppenheimer et al. "Urban Design Portfolio." *Architecture* (AIA, vol. 79, no. 4, April 1990): pp. 74–89 (Mission Bay).

Dean, Andrea Oppenheimer. "Profile: SOM, a Legend in Transition." *Architecture* (AIA, vol. 78, no. 2, February 1989): pp. 52–59.

Dillon, David et al. "Office Buildings." *Architecture* (AIA, special issue, vol. 77, no. 1, January, 1988): pp. 49–91 (Texas Commerce Tower, Dallas; Connecticut General, Bloomfield; Pacific Bell Administrative Center, Danville).

Dillon, David et al. "Preserving Modernism." *Architecture* (AIA, vol. 81, no. 11, November 1992): pp. 26–29, 61–99 (100 East Pratt Street, Baltimore).

Dixon, John Morris "Civics Lesson." *Progressive Architecture* (vol. 74, no. 6, June 1993): pp. 118–123.

Dorris, Virginia K. "Glass under Tension." *Architecture* (AIA, vol. 82, no. 9, September 1993): pp. 157–161.

Doubilet, Susan & Donkervoet Boles, Daralice. "Developers and Architects." *Progressive Architecture* (special issue, vol. 66, no. 7, July 1985): pp. 69–118 (LTV Center, Dallas).

Ellis, Charlotte et al. "American Work Abroad." *Architecture* (AIA, special issue, vol. 78, no. 1, January 1989): pp. 42–90 (Shell offices, The Hague).

Fisher, Thomas. "A Bow to Bahrain." *Progressive Architecture* (vol. 69, no. 5, May 1988): pp. 65–73 (United Gulf Bank).

"Full of East End Promise?" *Landscape Design* (no. 192, July/August 1990): pp. 8–9 (Canary Wharf).

Gaskie, Margaret F. "Building Types Study: 622. Urban Infill. Appropriate Architecture." *Architectural Record* (vol. 174, no. 1, January 1986): pp. 91–105 (Grand Hotel and Office Building, Washington, DC).

Gaskie, Margaret F. "Chicago Style: 303 West Madison Street, Chicago, Illinois." *Architectural Record* (vol. 177, no. 10 (9), September 1989): pp. 92–95.

Graham, Bruce. "The Architecture of Skidmore Owings and Merrill." *Royal Institute of British Architects. Transactions* (vol. 2, no. 2 (4), 1983): pp. 76–84.

Gunts, Edward et al. "Corporate Workplaces." *Architecture* (AIA, vol. 82, no. 10, October 1993): pp. 41–79.

Harriman, Marc S. "Breaking the Mold." *Architecture* (AIA, vol. 80, no. 1, January 1991): pp. 77–83 (Aurora Municipal Justice Center).

Harriman, Marc S. "London Bridge." *Architecture* (AIA, vol. 79, no. 9, September 1990): pp. 109–112 (Exchange House, Broadgate).

Harrington, Kevin et al. "Chicago." *Inland Architect* (vol. 37, no. 3, May/June 1993): pp. 42–59.

Heck, Sandy et al. "Interiors." *Architecture* (AIA, special issue, vol. 77, no. 6, June 1988): pp. 53–93 (Three railway station renovations in the Northeast).

Heck, Sandy. "New Scaled-down Scheme Unveiled for Coliseum Site." *Architecture* (AIA, vol. 77, no. 7, July 1988): pp. 38–39.

Heck, Sandy. "SOM/NY Replaces Safdie for Coliseum Site Development." *Architecture* (AIA, vol. 77, no. 2, February, 1988): pp. 13, 17, 19, 21, 23, 25.

Henderson, Justin, Truppin, Andrea & Jackson, Paula Rice. "Hotel/Restaurant Design." *Interiors* (vol. 148, no. 3, October 1988): pp. 128–143 (Rowes Wharf).

"Hopes Rise for Downsized Trump City Design." *Architectural Record* (vol. 179, no. 5, May 1991): p. 22 (Riverside South).

Jackson, Paula Rice et al. "14th Annual Interiors Awards." *Interiors* (vol. 152, no. 1, January 1993): pp. 79–117, 120, 126 (Contrax Office, Budapest).

Jackson, Paula Rice et al. "5th Annual Corporate America Design Awards." *Interiors* (vol. 152, no. 5, May 1993): pp. 111–147.

Kliment, Stephen A. "Manhattan Mosque: Islamic Cultural Center of New York." *Architectural Record* (vol. 180, no. 8, August 1992): pp. 90–97.

Krause, Cornelia et al. "Hoher als Breit." (High as Wide.) *Deutsche Bauzeitung* (vol. 127, no. 7, July 1993): pp. 13–73, 96–106, 112–119.

Kuma, Kengo et al. "The Third Wave in Skyscrapers: 22 Recent Super High-rise Office Buildings by SOM." *SD* (special issue, no. 276 (9), September 1987): pp. 5–48.

Leon, Pradeep. "Banking on Asia's Development." *Southeast Asia Building* (August 1989): pp. 26–36 (Asian Development Bank Headquarters, Manila).

Linn, Charles. "Two Faces Forward: Spiegel Corporate Headquarters, Downers Grove, IL." *Architectural Record* (vol. 181, no. 7, July 1993): pp. 68–73.

Lueder, Christoph. "Exchange House." *Deutsche Bauzeitschrift* (vol. 41, no. 7, July 1993): pp. 1189–1194.

Lyndon, Donlyn. "AIA Honor Awards: Distinction in Place." *Architecture* (AIA, vol. 83, no. 5, May 1994): pp. 105–123.

Lyndon, Donlyn. "The Station and the Wharf, Boston." *Places* (vol. 5, no. 4, 1988): pp. 3–12.

Magnusson, Emanuela. "In un interno di SOM a New York: il 'Palio' di Sandro Chia." (In an SOM interior in New York: the Palio murals by Sandra Chia.) *Domus* (no. 683, May 1987): pp. 6–8.

Mays, Vernon. "Light for the Site." *Progressive Architecture* (vol. 69, no. 12 (11), November 1988): pp. 108–113 (Pacific Bell Administration, San Ramon).

Mazzara, Gabriele. "Bank Headquarters, Jeddah, Saudi Arabia." *Industria delle Costruzioni* (vol. 18, no. 155, September 1984): pp. 46–51.

McKee, Bradford. "Urban Competition Held for Ho Chi Minh City." *Architecture* (AIA, vol. 82, no. 12, December 1993): pp. 28–29.

Melvin, Jeremy et al. "Flat Roofing." *Building Design* (no. 994, July 13, 1990): pp. 30–38 (Broadgate).

Melvin, Jeremy. "Digging a Life after Coal." *Building Design* (no. 1176, June 10, 1994): pp. 16–17.

Melvin, Jeremy. "Learning at Ludgate." *Building Design* (no. 1094, September 18, 1992): pp. 20–21.

Mickenberg, David. "Of Art and Architecture." *Inland Architect* (vol. 31, no. 5, September/October 1987): pp. 15–20 (Art Institute of Chicago).

"Mixed-use High Rise in San Francisco." *Progressive Architecture* (vol. 64, no. 1, January 1983): pp. 114–115.

Morganti, Renato. "Edilizia Ospedaliera a New York." (Hospital Buildings in New York.) *Industria delle Costruzioni* (vol. 26, no. 251, September 1992): pp. 38–49.

Murphy, Jim. "The Providence Connection: Providence Station, Providence, RI." *Progressive Architecture* (vol. 68, no. 3, March 1987): pp. 92–95.

Nesmith, Lynn et al. "The Tenth Annual Review of New American Architecture." *Architecture* (AIA, special issue, vol. 76, no. 5, May 1987): pp. 46–212 (National Commercial Bank, New York).

Newman, Morris. "LA Towers that Don't Forget the Pedestrian." *Progressive Architecture* (vol. 73, no. 12, December 1992): pp. 13–14 (Gas Company Tower).

"Northwest Frontier Province Agricultural University." *Progressive Architecture* (vol. 66, no. 1, January 1985): pp. 146–147.

Oliva, Caterina. "La Nuova Stazione Ferroviaria di Providence" (New Railway Station at Providence, Rhode Island.) *Industria delle Costruzioni* (vol. 22, no. 197, March 1988): pp. 40–43.

Pedersen, William et al. "On High-rise Residences." *Process: Architecture* (special issue, no. 64, January 1986): pp. 4–167 (New York Coliseum site).

Peters, Paulhans et al. "26 Platze." (26 squares.) *Baumeister* (special issue, vol. 86, no. 2, February 1989): pp. 11–53 (Canary Wharf).

Pousse, Jean-Francois et al. "Les Tours: Vers une Innovation Spatiale." (Towers: towards a spatial innovation.) *Techniques & Architecture* (no. 372, June/July 1987): pp. 46–118.

Powell, Ken & Jencks, Charles. "AD Profile 91. Post-Modern Triumphs in London." *Architectural Design* (vol. 61, no. 5/6, 1991): pp. 6–93 (Broadgate).

"Preview '86: 1 The Docks." *Architectural Review* (vol. 179, no. 1067, January 1986): pp. 20–30 (Canary Wharf, London).

Rabeneck, Andrew. "Broadgate and the Beaux Arts." *Architects' Journal* (vol. 192, no. 17, October, 24, 1990): pp. 36–41, 44–51.

Rawson, John "Working Details. Street Furniture: Broadgate." *Architects' Journal* (vol. 194, no. 11, September 11, 1991): pp. 49–51.

Ridout, Graham et al. "Canary Wharf: A Landmark in Construction." *Building* (special issue, vol. 256, no. 7717 (42) supplement, October 1991): pp. 4–114.

Ridout, Graham. "Arch Revival." *Building* (vol. 254, no. 7589 (11), March 17, 1989): pp. 39–44 (Broadgate).

Ridout, Graham. "Making a Million: Canary Wharf—the Next Phases." *Building* (vol. 256, no. 7717 (42), October. 18, 1991): pp. 21–23.

Russell, Beverly. "Interiors Awards." *Interiors* (vol. 147, no. 6, January 1988): pp. 138–173 (Palio Restaurant, New York).

Russell, Beverly. "Two at the Top." *Interiors* (vol. 151, no. 11, November 1992): pp. 54–57 (Credit Lyonnais, New York).

Sachner, Paul. "Building Types Study: 685. High-rise Office Buildings. High-risk High Rises." *Architectural Record* (vol. 178, no. 11 (10), October 1990): pp. 87–101 (AT&T Chicago).

Sachner, Paul. "Harboring Tradition: Rowes Wharf, Boston, Massachusetts." *Architectural Record* (vol. 176, no. 3, March 1988): pp. 86–93.

"Skidmore Owings & Merrill: Madison Plaza, Three First National Plaza, 33 West Monroe Building, 919 North Michigan Avenue, Park Avenue Plaza, Vista International Hotel, plus 5 projects." *A + U* (no. 3, 150, March 1983): pp. 23–58.

"SOM." *Tasarim* (no. 40, December 1993): pp. 106–109.

Spring, Martin. "Stretching City Limits." *Building* (vol. 253, no. 7569 (42), October 14, 1988): pp. 41–48 (Broadgate).

Spring, Martin. "Tea Party Spectacular." *Building* (vol. 253, no. 7532 (5), January 29, 1988): pp. 42–44 (Rowes Wharf).

Spring, Martin. "The Sky is the Limit: Rising above Their Station." *Building* (vol. 252, no. 7515 (39), September 25, 1987): pp. 51–61 (Broadgate).

Stern, Julie D. "Mission Bay Revisited." *Urban Land* (vol. 51, no. 4, April 1992): pp. 8–9.

Stewart, Alastair "Winning Entries." *Building* (vol. 256, no. 7684 (7), February 15, 1991): pp. 44–45 (Broadgate, Phase VII).

Stewart, Alastair. "Eighth Wonder of the World." *Building* (vol. 259, no. 7837 (13), April 1, 1994): pp. 14–15.

Stucchi, Silvano. "Due Recenti Realizzanioni del SOM: uno Spunto di Riflessione." (Two recent projects by Skidmore Owings & Merrill.) *Industria delle Costruzioni* (vol. 22, no. 206, December 1988) pp. 28–39 (Texas Commerce Tower, Rowes Wharf).

Swan, Russ. "Regards to Broadgate." *Concrete Quarterly* (Autumn 1989): pp. 8–14.

Swenarton, Mark. "Full Circle at Broadgate." *Building Design* (no. 875, March 4, 1988): pp. 14–16.

Taylor, Brian Brace et al. "Interiors for Public Use." *Mimar* (no. 16, April/June 1985): pp. 17–41 (National Commercial Bank, Jeddah).

"305 West Madison." *Progressive Architecture* (vol. 66, no. 1, January 1985): pp. 116–117.

Truppin, Andrea & Tetlow, Karin. "Trading Up." *Interiors* (vol. 147, no. 8, March 1988): pp. 153–188 (Merrill Lynch, New York).

Truppin, Andrea. "SOM at Fifty." *Interiors* (vol. 146, no. 5, December 1986): pp. 145–167.

"United Gulf Bank." *Progressive Architecture* (vol. 65, no. 1, January, 1984): pp. 104–105.

"Urban Focus." *Architecture Today* (no. 19, June 1991): pp. 57–58 (Cabot Square, Canary Wharf).

Wagner, Friedrich. "Making Light of Quality: John Lewis and Beyond." *Architects' Journal* (vol. 193, no. 15, April 10, 1991): pp. 51–54 (Broadgate).

Welsh, John "Tight Line." *Building Design* (no. 984, May 4, 1990): pp. 32–33 (Ludgate).

Whiteson, Leon. "Street Scenes." *Architecture* (AIA, vol. 80, no. 3, March 1991): pp. 108–115 (Columbia Savings & Loan).

Woodbridge, Sally B. "Mission Bay." *Progressive Architecture* (vol. 71, no. 5, May 1990): pp. 121–122.

Woodbridge, Sally. "A Set Piece." *Progressive Architecture* (vol. 68, no. 4, April 1987): pp. 108–112 (388 Market Street, San Francisco).

"Working Details. External Wall: Offices." *Architects' Journal* (vol. 192, no. 17, October 24, 1990): pp. 57–59 (Exchange House, Broadgate).

Zohlen, Gerwin et al. "Das Dickicht der Stadt." (The urban jungle.) *Baumeister* (special issue, vol. 88, no. 11, November 1991): pp. 11–67, 108–111 (Broadgate).

Acknowledgments

The partners would like to thank the following individuals for their contributions to SOM: Selected and Current Works. Paul Webber, Associate Director in the London office, co-ordinated the entire project, acting as liaison between SOM and The Images Publishing Group. Philip Mahla (SOM/New York) co-ordinated the collection and selection of photographs for each of the six SOM offices. Gretchen Bank (SOM/New York) contributed all project descriptions and, with Paul Webber, was responsible for final editorial review. Janet Male (SOM/London) worked on the final co-ordination of text and photos, as well as co-ordinating final revisions with Images. Deanna Kwan (SOM/New York) completed final color review on site in Australia at Images.

SOM photography co-ordinators from each office whose efforts are greatly appreciated include Maggie Diab (SOM/San Francisco); Karla Kaulfuss (SOM/Chicago); Deanna Kwan (SOM/New York); Janet Male (SOM/London); Naomi Miller (SOM/Washington, DC) and Tom Stangl (SOM/Los Angeles).

This publication also reflects the substantial influence of many other members of the design teams, including associated consultants, model makers, and SOM's dedicated and loyal multi-disciplinary staff.

Photography and Drawing Credits

Joe Aker: 110 (2); 111 (3); 118 (1); 119 (2, 3); 120 (4); 166 (1); 167 (2, 3); 168 (4); 169 (5, 6)

Steinkamp-Ballogg: 19 (3); 54 (1); 55 (2); 65 (5); 71 (3); 74 (1); 153 (2, 3); 215 (3); 226 (1); 227 (2)

Hedrich-Blessing: 57 (3); 66 (2); 67 (3); 70 (1, 2); 72 (1); 73 (2); 76 (5); 77 (6); 78 (7); 79 (8); 84 (1); 85 (2); 90 (2); 91 (3); 122 (1, 2); 156 (1); 157 (3); 158 (1, 2); 159 (3); 164 (1); 165 (2, 4); 175 (2, 3); 176 (1, 2); 177 (3–6); 178 (2); 179 (3–6); 212 (1); 218 (1); 219 (2, 3)

Hedrich-Blessing/Jon Miller: 123 (3); 124 (4, 5); 172 (1, 2); 173 (3); 187 (6)

Tony Brien: 139 (3); 140 (5)

Whitney Cox: 194 (1, 2); 195 (3, 4)

Alan Davidson: 104 (1); 105 (2)

John Donat: 139 (2)

ESTO: 63 (5)

ESTO/Peter Aaron: 50 (1); 51 (2); 130 (1); 131 (2); 132 (3); 133 (4, 5)

ESTO/Scott Frances: 53 (3)

ESTO/Brian Nolan: 83 (2); 93 (3)

ESTO/Jock Pottle: 32 (1); 33 (2); 41 (4); 103 (2, 3); 229 (3)

Simon Hazelgrove: 140 (4)

Wolfgang Hoyt: 48 (1); 49 (2–4); 60 (2); 61 (3); 62 (4); 86 (1); 87 (2, 3); 88 (4); 89 (5, 6); 126 (1, 2); 127 (3, 4); 134 (6); 135 (7, 8); 162 (1); 163 (2, 3); 180 (2); 181 (3); 182 (4–7); 183 (8); 189 (3–5); 197 (7); 200 (2); 201 (3)

Alistair Hunter: 154 (1); 155 (2, 3)

Timothy Hursley: 165 (3, 5)

Jane Lidz: 68 (2); 69 (3); 114 (2); 115 (3)

Heiner Leiska: 47 (3, 4)

Nathaniel Lieberman: 52 (2)

Michael McCann: 40 (1, 2); 41 (3); 151 (3, 4)

John Miller: 138 (1); 141 (6)

Jon Miller: 184 (2); 185 (3, 4); 186 (5)

Robert Miller: 190 (6, 7); 191 (8, 9)

Michael Moran: 128 (1); 129 (2–5)

James H Morris: 25 (3); 26 (4); 27 (5–8); 98 (1–3); 99 (4); 100 (5); 101 (6–8); 106 (2); 107 (3–6)

Gregory Murphy: 116 (1); 117 (2, 3)

Machiko Nakahira: 108 (2); 109 (3, 4)

Andrew Putler: 105 (3); 161 (5)

Richard Rochon: 35 (2); 38 (1); 39 (2); 82 (1); 92 (2); 102 (1); 142 (1); 143 (2)

Steve Rosenthal: 30 (4); 31 (5); 203 (2, 3); 204 (4); 205 (5–7)

Abby Sadin: 174 (1)

Richard Waite: 20 (5)

Nick Wheeler: 28 (2); 29 (3); 56 (2)

Alan Williams: 20 (4); 21 (6); 22 (7); 23 (8); 24 (2)

Roy Wright: 80 (2); 81 (3); 192 (2)

Index

Bold page numbers refer to projects included in Selected and Current Works.